BILL LISHMAN

Father Goose

LITTLE, BROWN AND COMPANY (CANADA) LIMITED
Boston • New York • Toronto • London

First published in hardcover by:
Little, Brown and Company (Canada) Limited 1995
This edition published by:
Little, Brown and Company (Canada) Limited 1996

Canadian Cataloguing in Publication Data

    Lishman, Bill
        Father Goose

ISBN 0–316–52708–4 (bound)
ISBN 0–316–52742–4 (pbk.)

1. Lishman, Bill.   2. Geese — Ontario — Nestleton.
3. Geese.   I. Title.

QL696.A52L58 1995   590.4'104525'092   C95–931439–3

Credits
Front jacket and interior photos: Joseph Duff
Interior design and electronic assembly: Pixel Graphics Inc.
Printed and bound in Canada by:
    Best Book Manufacturers Inc.

LITTLE, BROWN AND COMPANY (CANADA) LIMITED
148 Yorkville Avenue, Toronto, Ontario, Canada

DEDICATION

This book has to be dedicated to the memory of my mother Myra White Cronk who, from her humble beginnings as a Quaker farm girl at the start of this century, went on to be one of the few women of the era to earn a Master's degree in Biology from the University of Toronto.

She gave up her academic career to raise a family. She not only gave me my biological start but instilled in me a reverence for nature that has only become stronger over the years.

TABLE OF CONTENTS

ACKNOWLEDGEMENTS

A big thanks to Scott Young, Allan Herscovici, Owen Neill, and Christina Fox for their assistance in helping shepherd the words of this book together.

In terms of my personal inspiration and support, there's no one more deserving of my deepest gratitude than my family —— my dear wife, Paula, sons Aaron and Geordie, and my daughter, Carmen. Richard VanHeuvelen, who has worked with me for the past seventeen years, I can almost include as a member of my family. Without their enduring support, many things may have fallen by the wayside.

With regard to the great adventure of flying with birds and the two southward journeys, I have to acknowledge the tremendous input of Joseph Duff, William Carrick, and Dr. William Sladen.

There were a multitude of people that became inspired by the great bird-flight adventure who bent over backwards to help out at one critical juncture or another. The list could fill pages but the reader will discover most of them as the book unfolds. During our flights south, I was always overwhelmed by the co-operation and hospitality shown by all. From Dorothy and Frenchy Downey to Jim and Sandy Compton; Miss Peggy Augustus to Bob and Cornelia Joyner, and a host more.

A special thanks to Bruce Westwood of Westwood Creative Artists Ltd. who became so involved in the project that he volunteered to be a founding board member of *Operation Migration*.

Thank you!

Most of the photography, particularly the in-flight pictures, can be credited to Joseph Duff.

BY OWEN NEILL

Goose Send-Off

The voice of ancient gossamer
rises still its ancient way
in the quiet thunder of eager wings
trying that the north winds say.

   What do winds whisper in clever ears
   that makes the time to go just right?
   The wonder is the message comes
   to all at once like second sight.

Preparation, patient, strong
born from the egg with what they need
each bird follows its primal plan
ten thousand years could not impede.

   We came, once, with reckless sway
   and cut across the natural flow.
   But now repentance pricks us on
   as we repair the status quo.

We set the wild geese wild once more.
We hatch, imprint and train the flock.
We run, then fly, again, again
until we share the same bloodstock.

Man and bird are strangely one
yet each cannot know the other.
Along the route soon remembered well
the wild and tame are somehow brother.

The goal is fixed in the mind as one.
All eyes peel the horizon away.
It's time to soar where long ago
all nature held its solemn sway.

Motion liquid below autumn clouds
a man-bird combination flies.
Like an ancient myth we integrate
and put sweet mystery in our eyes.

A wrong is righted heroically.
A beauty we have never known
gives hope our world at least in part
will reap the harvest we have modestly sown.

The voice of ancient gossamer
rises again its ancient way
in the quiet mystery of newborn wings.
They know again what the north winds say.

Prologue

At about ten to seven on a Sunday morning, September 2, 1990, Clive Beddall's station wagon rolls quietly down my drive. The tops of the forty-foot poplars south of the house indicate there is a slight breeze out of the west. The sun has risen clearly and there are again those little patches of mist in the hollows of the valley.

We grab a quick coffee and formulate our flight plan. Clive and Tim Allen will fly the Cosmos two-place, Tim with the video camera. The plan is to let our main flock of nine geese out for their warm-up flight on their own. When they return, I will take them off with the Riser. Clive and Tim will follow in the Cosmos, hoping to get into formation for video footage. I take the truck to the shop and pick up the camera and a freshly charged battery while Clive and Tim preflight the Cosmos. When I return to the strip, I let the geese out and they take off immediately and fly low level, silhouetted into the rising sun. As I stroll down to the hangar, Tim calls for my attention, pointing north. For a moment I mistakenly think it's my flight of geese returning already, but no. There are about twenty, no, thirty geese approaching low from the northwest. Then, hearing a honking to the east, I catch sight of my nine flyers coming back above the runway. They pass a few hundred feet in front of the wild flock and, to our astonishment, the wild flock executes a full ninety-degree turn and in a flash they have formed up with my nine. There is a flurry of honking and my group dissolves into the ragged V. Within seconds the whole skein disappears from view over the horizon to the south. My heart almost stops. Tim and Clive and I stand there momentarily riveted — as if we had just watched our children kidnapped before our eyes.

My hesitation was short-lived. I wheeled out the Riser and gave it a thirty-second preflight. Fortunately I had fuelled it the night before. Fuel on. Ignition on. Full choke. Starter button, pop! pop! gurgle! baroom! I ease back the choke. It smooths out to an idle. I give it another twenty seconds to warm up — twenty seconds that seem so long I could have read the Sunday paper cover to cover — as my geese slip away to the south, captured by their wild cousins, honking and gabbing as they distance themselves from their home. They had to be at least two miles away by then. Full throttle, wet grass slipping by, airspeed reading twenty-five miles per hour, back on the stick. The grass drops away. By the time I climb out in a tight curve, barely clearing the poplars to the south, they are nowhere to be seen. Many questions pass through my mind as I level off at about five hundred feet. Have I lost them permanently? Will they just fly a little bit with their wild cousins and circle back? Will they bring the whole flight back with them? Will I be able to catch up to them in the air? If so, what then?

The air was steady, the bright morning sun glistening off the trees that protruded out of the slight morning mist. I maintained full throttle power and trimmed the Riser for full speed. The airspeed was reading almost fifty miles per hour, the Riser's cables singing. If the geese were cruising at their customary thirty, could I catch them before I had to turn back at the Oshawa Airport control zone? Had they set down in one of the many ponds to the south? Had they changed direction? I kept my southward heading at five hundred feet, the air rock-steady with about a ten-mile-an-hour wind drifting me to the east.

Crossing the ridge of forest that stretches for miles to the east and west, I strained my eyes in all directions for any sign of movement. It seemed a hopeless quest. Perhaps I should just turn around and head back. I passed over abandoned gravel pits where a few dirt bikers were unloading, their machines fluorescent against the dull grey of the dunes. I flashed back to the days when that was me getting ready to ride my antiquated Triumph. I was ready to yell down and ask if they had seen the flight of geese but understood the futility of that idea.

South of the ridge the land breaks up into rolling farmland creased with meandering wooded valleys, dotted with numerous ponds. They could have landed in any of those ponds. Or they could have picked a stubble field.

I was approaching the boundary of the Oshawa Airport control zone, banking to the west. I looked for a large pond that I knew wild geese frequented, simultaneously glancing north to see if the boys in the Cosmos were airborne. There was no sign of its hang-glider form… but wait! Low down about a mile off to the north, I saw some movement, a tiny string of motion, a beaded thread of flashing wingtips in the morning sun against the darkness of the evergreen ridge. Stick full right, I cranked the Riser around into an intercept heading. The wind meter on the second strut read 50 mph. The geese were at low level, maybe a few dozen feet off the treetops.

There was a power line between us, the old rusty towers barely discernible and the wires invisible, deadly spider strands, so I maintained altitude until I was past the power lines, then dropped down to intercept the flock.

There must have been about fifty geese, beautifully lit by that dawn sun against the dark pine forest.

Oh please, will my nine break out and come home with me or are they gone forever, enthraled, their instincts awakened to their true environment? I was closing fast. At about three hundred yards the flight leader saw me and executed a tight, sweeping 180-degree turn to the west. Now I was in a trailing position. Standing the Riser on its wing, I bent it around to follow. The flight now was veering to the south — my nine must be in there — if I could only catch them up maybe, just maybe, they would break out and form up with me. Still at full throttle I tried to outflank their southerly heading. They were low, perhaps 150 feet. The westerly breeze had picked up a little and at the low level the air became rough. Those power lines were out there somewhere. I bounced through the valleys. I glimpsed one quaint farmstead nestled in a valley and caught the name on a parked van: "Ken Jen Petting Zoo". I was now flying almost abreast and to the south of the main flock paralleling the ridge. Which are mine and why aren't they breaking out to

Bill Lishman

form up with me? Maybe those guys from the Wildlife Service are right after all. Perhaps they now have escaped into the wild. I had to head off the leader. I could not take the chance of them getting away. The flying took all my concentration, like herding wayward cattle except that in this case there was no fence but the fields and woods below.

Over my shoulder at last, a mile distant, I finally caught sight of the Cosmos silhouetted in the easterly sky. I wished that we had radio on it, working.

The lead goose veered the flight to the north. We were approaching the ridge again. They didn't climb, but let the forest rise to meet them. They crossed over just west of a skeet range, some of the geese actually flying through the treetops. I was down there too, treetops passing ten or twenty feet beneath me. The turbulence off them kept me working to keep the Riser level.

I moved up on the leaders again. If I could veer them farther east, my birds, which were probably reaching their limit of endurance, might drop out. At that point I got really close to the whole flight. Now they were within fifty feet of me. None broke away. A mass of about twenty were in the lead, the rest strung out in a long straight line. We held this position for a while and I had a chance to relax and enjoy the beauty of this amazing flight of creatures, like a school of fish passing over a reef, that timeless fluid image. If I could maintain them on this heading, it would bring them right back over our home at Purple Hill and the airstrip. If they broke away to the southeast I would lose them again. These birds were fast, at least ten miles an hour faster than what I was used to. The leaders were flapping like crazy ducks, the guys in the back almost gliding.

Then we were over six thousand acres of forest that borders our property to the west and north. It was obvious they wanted to head north into the marshes of Lake Scugog. I glanced around for Clive and Tim in the Cosmos. They were still off to the south, trying to catch up. It was a treetop struggle. The geese kept wanting to veer north. At that altitude I was really being bounced and felt the first queasiness of motion sickness.

In my haste at take-off I had left my watch behind. How long had I been flying at full throttle? How much fuel did I have left? I tried to see what was left in the tank, but in that rough air I could not.

The leaders had broken away to the north again. I concentrated on getting them herded farther east. We crossed back over the north boundary of my property. I was on the northerly periphery of the flock. I managed to get them headed south right toward our airstrip. Surely my geese would drop out… but just then the flight broke into half a dozen independent groups, all heading for the marsh again. I cranked the Riser around in another wing-stander and caught a glimpse of Clive in the Cosmos just now catching the tail-end of the action.

I picked one flight to follow. It looked like about nine geese and I tried to catch up. My stomach was churning from motion sickness, worry about fuel and fear that my whole plan had been blown by an unlucky chance. I broke off the pursuit, throttled back, and headed for home. I met the boys in the Cosmos and pointed at my fuel tank. They kept heading in the general direction of the way the geese had gone.

As I banked right for final approach, I saw something

Bill Lishman

like small dots on my runway. There were nine geese nonchalantly grazing in front of the hangar.

I set the Riser down and rolled up to their noisy greeting, as if they were asking where had I been.

I was asking too. Were they ever in that group I followed? Did I miss them breaking out and heading home? Or had I just been on one more wild-goose chase?

Anyway, my nine were home, back on the ground. They didn't fly off into the wild after all. They just had a wild interlude.

CHAPTER ONE

Woodlawn Farm

When I am asked how I wound up in mid-life as one of the principals of a movement that might eventually have a profound effect on preserving some species of birds from extinction, the honest answer has to take in a good deal of territory. I think it began when I was a boy not yet in my teens, rafting on a two-acre pond that was part of my father's cattle farm in south-central Ontario. He had created the pond by damming up a creek that meandered through several deep pools that had been our swimming hole. These pools teemed with life. My mother, who was an expert in the fauna of freshwater streams, identified many for me. There were stickleback, chub, catfish, leeches, crayfish, clams, and many more species. She held her Master of Arts degree in biology from the University of Toronto. She had taught there until she married and moved back to her family's ancestral farm. Her love of the natural world was passed on to me at an early age. She spent many hours talking to me about everything from the teaching of Darwin to the inner workings of a chicken as we cleaned/dissected it for Sunday dinner. My father taught me a lot about our farm machinery and how a boy could make things that worked as well as machines.

The pond attracted many wild waterfowl, especially mallard ducks. I made a raft from old railway ties and fence rails lashed together with baler twine and sat on it hidden among the reeds, watching the ducks and other water creatures come and go.

I practised mimicking the mallards' call. One morning, when the fields had been plowed and the last vestiges of leaves had blown from the trees, I spied a huge flight of ducks coming from the north. I tried my call once, then again and again. They turned and slowly circled the pond and finally landed. I was thrilled. I had learned a little of their language.

I made a small water wheel that turned merrily in the flowing water of the creek. I also carved wooden shapes and held them in the flow. By stirring up a little mud upstream from the pieces, I watched how the brown water flowed about the shapes. In this simple manner I learned about the nature of airfoils, which helped me to understand fluid dynamics. There was one particular pool that fostered hordes of minnows. I would net these, put them in a pail, and see how many different kinds there were, varied and so colourful, always something new to discover.

There was a blue heron that hung out at the pond. He had two favourite areas where he would stand so still that sometimes you could not see him. He would wait for minnows or frogs to happen by and in a flash his stillness would turn to lightning speed as his beak speared his unsuspecting prey. One year a grebe nested at the pond and wood ducks made a nest in an old hemlock that the pond had surrounded. In the fall they all departed and again I was curious. Where did these birds go? What adventures did they have? I hoped that some day I could fly with them.

We had domesticated geese on the farm, and I learned something from them as well. George, a large mean grey gander, was the chief goose on the farm. His two wives were "Whitey", and a grey called "Nothing." They would raise a gaggle of goslings each year, some of which would become Christmas dinner for us and the neighbours. These geese were free to roam the farm. My father claimed that, with the noise they made when disturbed, they were better than a watchdog. George seemed to know this. He strode in a haughty air about the farmyard and nobody messed with him. If you got too near the flock, he would herd the geese behind him, then turn and spread his mighty set of wings and hiss defiantly. If you persisted, he would rush at you with great ferocity. Only a fool would stand up to his onslaught, for a rap from a gander's wing can and has broken bones.

Whitey, my favourite, was a gentle goose who almost let you pet her on the rare occasions when George was not around. But geese, like most birds and unlike many domestic animals, do not like to be touched. To fly efficiently all feathers must be in place, so why would they risk some outsider disturbing their carefully manicured airframe?

The pond was downhill from the barn. In between there was an ancient pear orchard planted by one of my forefathers. Not the modern dwarf trees, but trees thirty and forty feet high. The cattle grazed beneath them.

Each day the geese would waddle down from the barn through the pear orchard for a daily dip in the pond, all except Whitey, that is, for she could fly... almost. She would run down from the barn, flapping furiously, and get a few feet off the ground, mainly because it was a down-grade. Barely clearing the barnyard fence, she would weave wildly through the pear trees and cows until she tumbled into the pond with a great splash.

She never did learn how to get her twenty-some odd pounds onto that pond with any grace. Having arrived at the pond long before her earth bound family, she would desperately call out to them. Often in the morning as we brought the cows in for milking, she would make this awkward flight, and if you were in the wrong place you took a chance of being flattened by a low-level kamikaze goose. Sometimes she would startle the cows and they would stampede.

Each fall and again in the spring the whole flock would attempt to take off. Years of domestic breeding had removed their ability to fly, but the instinct to migrate could not be removed. Whether it was a change in the length of the day, a snap in the weather or whatever tripped that migratory instinct, an excitement would spread among the earthbound flock, like schoolkids anticipating the coming of summer holidays. For several days there would be much flapping of wings, as a pilot might run up the engine on his aircraft, or an athlete go through warm-up exercises. The geese would stretch their wings, then do a few-seconds flap, generating a cloud of dust and feathers across the yard. Much excited gabbling took place among the flock. This noise would build until they would all gather at the top of the barn ramp. Then, with George in the lead, Whitey and the yearling goslings strung out behind, they would make a frenzied run down the barn ramp and across the barnyard, a great excited flurry of an attempt at take-off, but only Whitey could get airborne. She would, in her usual fashion, barely clear the fence for her thirty-second flight, weaving through the pear trees and the cows to end in a great splash in the pond. Time and again they would try to get airborne for some imagined flight south. Whitey would waddle back up through the orchard calling to the flock. They would await her return and then there would be a great flurry, heads bobbing up and down as if they were discussing what went wrong and what take-off procedure they should try next. Once again they would build up to another attempt. I could feel their frustration, for their timing coincided with that of the wild geese high above, honking their way either north or south in great chevrons. To see these wild ones was always a thrill. I felt that our domesticated geese, earthbound, were envious of the freedom these wild birds had. It stirred in me a sense of wonder and adventure: Where did they come from? Where were they going? What map within them charts the way? How do they know when to fly? I longed to be up there with them to see the world from their vantage and to discover the places where they went, but like our domestic geese it seemed I was relegated to stay on the ground and only dream about their southbound adventure.

Bill Lishman

CHAPTER TWO

Wanting to Fly

One day in the late forties, a light plane came swooping in over our fields, skimming along inches above our crop of peas, spewing a fog of insecticide. It flew over one set of wires, under the high-tension lines between the pylons, and then pulled up — missing the barn by inches. I stood out there at the end of the field on its second run watching as it sped past at seventy or eighty miles an hour. I was eight years old.

After that I *had* to fly. I wasn't yet fourteen when I joined the Royal Canadian Air Cadets 151 Squadron in Oshawa. Once a week, whatever the weather, I would put on the air cadet uniform, polish the brass buttons, and centre my hat two fingers above the eye, according to regulations. Sometimes I even spit-polished my shoes. I would then walk or bicycle the gravel mile down to Highway 2, hitchhike ten miles into Oshawa, walk another seven blocks, and then usually spend another hour or so marching about in the creaky old wooden Rotary Hall that was reserved for the air cadets on

Tuesday nights. When drilling was over, we would take a course in something to do with flying, like navigation, or aero engines or meteorology. We learned about rhumb lines (constant compass direction), great circles, radial engines, altocumulus clouds, and all sorts of things that a good flier should know. When class ended, we would form up again and march some more before falling out. I would then reverse my trek and, if lucky, would arrive home sometime before midnight.

After two years of this routine plus two disastrous attempts at summer air cadet camp (one spent in a military hospital having my appendix removed and the other picking up garbage on the air base at Clinton, Ontario, for a minor disciplinary infraction), I had had only twenty minutes flying time, and that was as a passenger! My interest in air cadets as a means to get flying began to wane. I did, however, pass my ground-school exam, making me eligible for flight training — only to find that I was partially colour-blind, wiping out that chance for a flying scholarship. Still, I was sure I'd get into the air somehow.

When I was seventeen, my dad, needing extra help on the farm, hired an Ojibwa boy named Lee Antoine. Lee was "on sabbatical" from the Bowmanville Boys' Training School, or reform school as we called it in those days. He was just a bit over five feet tall, had a deadly aim with anything, and was as tough as well-gristled steak. He loved to fight and would spend hours practising his punch. When he flung his fist out like a whip, you could hear all his arm joints crack. He and I never had reason to fight, but I'd seen him in action. We became good friends.

One evening as Lee and I chatted at the dinner table, my mother became irritated at the important teenage stuff we were talking about, whatever it was. An argument ensued, as happens between teenager and parent. This turned into a yelling match, soon including my father, at which point I decided (at the speed of light) to leave both home and school and join the air force. I was sure my years in the air cadets would help. I was also convinced that the Royal Canadian Air Force was desperate for personnel of my calibre. I trudged off into the darkness to join the air force. That side-road mile to Highway 2 was never so long, and I will never forget that great ripping feeling within as I distanced myself from the farm. After an uncomfortable night with a friend in Ajax, I returned home the next day to pick up a few things, including a suit of Lee Antoine's that had been his official graduation garb at reform school, and which was too big for me. Then I caught the Toronto-bound Gray Coach bus for a three o'clock appointment with an RCAF recruiting officer.

I arrived hours early and drifted around downtown, stopping at a street vendor where I bought a travel alarm clock for five dollars, taking care to set it for a few minutes to three P.M. so I wouldn't miss my RCAF appointment. Being extra careful, I got there ten minutes early. I guess I was a rather arresting sight: a lanky 120-pound tiger dressed in a reform-school suit made for a 160-pounder, to say nothing of the loud ticking from one pocket, which somehow I didn't notice at all.

I stammered out my name to the receptionist who, to my surprise, ushered me right into the private office of a real uniformed RCAF recruiting officer. He sat behind a clean, organized oak desk, his uniform spotless, his hair in perfect order, the image one might see on a recruiting poster. I tried to straighten my windblown hair, and while he looked over my application I carefully polished the toes of my shoes on the back of my trousers (a trick learned while waiting for inspection as an air cadet). I don't recall much of the conversation. I kept trying to give intelligent answers, only to get myself hung up in blurted sentences that I couldn't end properly. I was just trying to extricate myself from one of these when a loud ringing came from my chest. I fumbled the jangling clock out of my inner pocket, caught it in free fall before it hit the floor, then spent another harried moment trying to open it up to get at the minuscule shut-off lever. Finally after it seemed to have jangled on for ten minutes, all was silent. The officer just stared, then quietly shuffled my application under a few other papers on his desk. Needless to say, I never became a fighter pilot and that was the end of my military career.

My final, or near final, break with my family happened when I had been living with my grandmother and had come to my parents' home for dinner one night. I made the mistake of sitting in a chair normally reserved for Lee. This angered my mother, who had designated one chair, the one I was using, for Lee because he normally smelled of the farm and she wanted him to use only that one chair. When I protested that treatment of Lee, a hot argument began with my mother. Lee left in a huff, and when my father came in and found me arguing with my mother he joined her in various denunciations of my activities. It turned into a free-for-all that sent me out of the house and into a boardinghouse in the nearby town.

My older sister, Alaine, had had her own battle with my father. My younger sister, Louise, sided with her and they left together. My father at that point didn't really know how to handle teenagers — if anybody does — but the result was that all three of us left home, my sisters going to Sault Ste. Marie to work, while I went back to my buddies at school.

One of my private projects at school was making gunpowder, an experiment that eventually led to a very large explosion that ruined a gravel company's weigh scale — an escapade I had opposed and was not involved in. Nevertheless, several students, including me were charged with a variety of explosive-related misdemeanours. The charge against me was dropped, but the school principal, an understanding man, suggested to me that the unrest in the school over this whole matter would be lessened if I left for a while. I joined my sisters in Sault Ste. Marie and finished the school year there.

In order to support myself, I took a job repairing outboards. When the demand for that line of work ended with the fall, I worked in construction on a labour gang digging ditches, until I heard of a motorcycle-repairing job in Cornwall, halfway across the province. But once there I found there wasn't enough work to keep me busy. I then made what turned out to be a major decision in my life — I enrolled in an art-school night course. The instructor was enthusiastic about my drawing and paintings, but I was restless and decided to move on again. I had bought a Triumph motorcycle in the Sault, which I sold for $400 and bought a one-way ticket to England on the *Queen Mary* ocean liner for $180.

I hitchhiked all over England, seeing a lot and often

staying with relatives. I discovered that my paternal grandfather, had been a wood-carver in Hexham, Northumberland, and had a whole set of wood-carving tools, which I found and claimed. I spent many hours sitting in wood-carving shops, watching and learning much from these old traditional English craftsmen. They told me where to buy chisels and how to sharpen them and what kind of woods to carve. They were solitary workers but seemed happy to share their knowledge with a young Canadian. I was having fun, but I ran very short of money and knew that I had to head back to Canada. An aunt in London, lent me forty dollars, a lot of money in those days, and my dad, who had some connections in the shipping business around Newcastle, helped get me passage on a grain boat headed for Canada.

The boat was an ancient wartime Liberty ship, in bad shape. Just west of Ireland we got into high winds and huge seas. Every time the bow came out of a wave, water spurted out of rivet holes. It was scary. We got blown away up around Iceland. Finally the sea calmed down a little and we plodded along through a whole field of icebergs, a wonderful experience in itself. As the only passenger, I learned how to steer the ship and how to navigate using the stars and the sun. The radio operator and I became good friends and he taught me something of his trade. I learned a lot coming across on that old ship before we sailed past Newfoundland and up the St. Lawrence River. After twenty-two days at sea, when we reached Lévis, Quebec, the boat was in such bad shape that it had to be put into dry dock immediately. With $1.50 in my pocket, I hitchhiked home the five hundred miles from Lévis to Toronto.

Bill Lishman

That winter, I enrolled in the foundation course at the Ontario College of Art. I did very well in art school, at least partly because of what I had learned of my grandfather's trade in England. The sculpture teacher said I was the best sculptor he'd ever had in school. I had high marks but I was so short of money that I was lonely. I couldn't date anybody or go out with the gang. Another drawback was that my father was not pleased at all about my interest in art; said he didn't have the money to support me; thought I should be doing something more practical, such as engineering.

When I got home from art school the following summer, Dad, in a sense, got even. He gave me a shovel and for a month and a half I worked on the farm's manure pile. Talk about Augean stables! That's the Greek legend about the king who kept three thousand oxen whose stables hadn't been cleaned for thirty years. Hercules cleaned these out in a single day by diverting the river Alpheus through the stables. I had no river to fall back on!

My next job was at a company now known as Dowty Aerospace in Ajax, not far from Pickering. I was office boy, salesman, public relations man, and demonstration driver for the boats that Dowty was making in the late 1950s while the aerospace business was in a slump. It was that summer that I decided to become a serious artist, but it was not until a chance encounter sometime later that I really set about it.

CHAPTER THREE

A Meaningful Pattern

In the early 1960s I met an old German who professed to be an expert palm reader. During World War II, he said, he had done statistical research based on reading the palms of soldiers going to war. What effect heavy artillery and other weapons of war would have on his calculations, he didn't say, but after convincing me that he could read palms with great accuracy, he told me that I would not live past thirty-eight. He certainly got my attention with that prediction.

I was twenty-two. My life had very little direction. I had been procrastinating over the years, figuring I had an eternity to do the things that I had daydreamed about. Generally I was lazy, following whims rather than setting goals. That episode with the old German really motivated me, convincing me that my life was finite, not only finite, but too short for comfort!

My interests had always been diverse. I truly wanted to do everything and had difficulty maintaining a single direction. The career philosophy of the day was to pick a field and become a specialist. That was not for me. The idea of getting boxed into one course grated on me relentlessly. Still, sculpture was my passion. I loved to carve and I had an instinctive confidence that I could literally make anything three-dimensional, and that three-dimensional design had to be my main direction. I had dreams of great things I wanted to build. To fulfil them, I needed to be able to do a number of things well — and since I was a terrible student in most classroom situations — where I could never stay awake — I tended to learn on my own. So my technique of learning was mostly by trying different things; jump in and go for it. It may have been the long way round to becoming a sculptor, but it certainly wasn't boring. I found that I could learn in the strangest of places. When I couldn't find answers in a book, or from someone who could show me, I invented techniques. In most instances I learned by trial and error.

Then I got lucky. One day I happened by a hundred-year-old blacksmith's shop in the village of Greenwood, Ontario. I used to visit the shop when I was a boy. The blacksmith had died, and I asked the woman who had inherited it if I could rent it as a studio. She was a fine woman. When I told her of my hopes of becoming a sculptor, she said, "I'll give it to you for a year, for free."

That opened a lot of doors for me, both actual and in my head.

Walter Wilson had been the blacksmith in Greenwood for more than fifty years and, aside from shoeing hundreds of horses, he built exceptional wagons, sleighs, and hand tools; there was not much Walter could not make. He died on the job in 1958 and the doors of the shop were closed, not to be reopened again until a good artist, Brian MacKenzie, an old high-school buddy, and I came along four years later. Moving in was exciting. It was like an archaeological dig or like opening a time capsule. Walter had had little use for modern machinery and so had accumulated a wealth of tools over the years. With the exception of primitive electric lighting and the huge old electric motor that powered the overhead line shafts, the equipment might well have been out of biblical times. Pinned up on the walls beneath decades of dust and soot were posters from both world wars and bills dating back to the turn of the century. The cobwebbed walls of the woodworking shop in the rear were hung with thin oak patterns for wagon and sleigh parts (and some patterns that we could never figure out). The place had a mellowness and a feel to it that only time can produce. We enjoyed every second of discovery.

We cleaned up most of the junk, and I started carving many kinds of wood — redwood off-cuts from a nearby sash company, apple, butternut, cherry, oak. I was fascinated with metalworking, too, and I cleaned up the forge, got it working, and made a lot of metal sculpture. It didn't take long for Walter's old cronies to filter back in through the reopened doors to take up their old seats and watch the two greenhorns attempting to bask iron and chip wood, with occasional interjections of "You're not doing that right... Walter used to do it this way..." They regaled us with humorous stories of past events in the shop and offered many examples of the gospel of blacksmithing according to Walter. On occasion as I looked at the weathered faces in the flickering dim light of the forge, I was sure Walter's ghost was there, prompting his old buddies to give us direction.

1

2

3

1   7 a.m. plane and geese.

2   Geese over forest.

3   Bill as a youth.

4   Bill (centre) in his blacksmith's shop with friends Mike Roche (left), and Bruce Lloyd (right).

5   *Horse* at Toronto City Hall, August 1967.

4

5

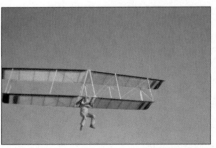

6　　The Lunar Lander.

7　　1978 foot launch of
　　　the powered Easy
　　　Riser.

8, 9　Airborne in the Easy
　　　　Riser hang glider.

10

11

13

12

10 Paula Lishman with the Easy Riser on frozen Lake Scugog.

11 Inspecting Purple Hill on a June morning.

12 Bill Carrick with goslings, exercising his geese by boat.

13 Bill Lishman with goslings.

14 A gosling.

15 Joe wears the "goose toddler".

16 Hatching eggs.

17 Bill and the geese go for a swim.

18

19

20

Murray Skuce

21

18　Geese in the tub.

19　Bill's son, Aaron, riding
　　with the geese pre-flight.

20　Bill and daughter
　　Carmen with geese and
　　the Easy Riser.

21　The Easy Riser with
　　landing gear.

22

23

24

25

Stephen Frank

22 Geese watching
   the Easy Riser.

23 Bill flies in for-
   mation with the
   geese, 1988.

24 Geese in the
   mist.

25 *Autohenge,*
   August 1986.

26

27

28

29

26 Taking a swim
    with some cygnets.

27 Bill and Joe Duff
    greet the geese at
    their landing in
    Virginia.

28 Ghostlike wire
    frame of our dome
    house under
    construction, 1990.

29 Aerial view of the
    house under con-
    struction, 1990.

30

31

32

33

30 Photograph of a
   whooping crane.

31 Geese over a cor-
   duroy field.

32 Southbound over
   New York state with
   thirty-eight geese in
   tow.

33 *Pregnant Woman* on
   delivery, 1993.

We eked out a living trying to sell our work on weekends. The general store gave us credit, and when we sold something we'd pay them back. To earn what little money we did make, we chipped out small souvenir carvings that we either sold out the door to tourists or tried to peddle in the Toronto craft shops. There was no walking into the bank manager's office and getting a loan. We became very adept at barter, and wily in methods of keeping Ontario Hydro and the telephone company from switching us off. As I had quit a "good job" at Dowty in Ajax to do this, my parents thought me totally foolish. In fact it took years for my dad, a "practical" man, to accept any different interpretation. No wonder. The first year, 1963, we grossed eight hundred dollars out of the business. But from then on things got better fast.

I did my first real metal sculpture in the summer of 1963, a horse made from pieces of a scrap Oldsmobile hood, welded together with an arc welder I made by mating an old Austin engine to an aircraft generator. A designer of commercial interiors, and a neighbour, bought the piece, and a month or so later asked me to participate in a design contest with his group. My design won, leading to my first real commission, nine pieces of wall sculpture in wood and metal for the flagship store of Canada's then largest grocery chain in Yorkdale Shopping Centre — at the time the most prestigious shopping centre in Canada.

Wow! I thought I had it made, but I can still recall the nervousness. It's one thing to get the job and another to figure out how to get it done and installed on time. I ran around in circles for a couple of weeks, discarding sketch after sketch until I had some semblance of order. I knew I needed help, in addition to Brian, to get this job done, so I hired two of our buddies (kindred misfits) who were hanging out at the blacksmith shop anyway.

Along with the rent-free (for a year) shop, we had rented the blacksmith's house, a lovely old board-and-batten two-storey primitive, a stone's throw east of the shop, nestled into the bottom of the hill. It needed paint and had a broken-down front porch.

When Brian and I moved in, we did fix it up a little. There was only a kitchen wood stove to keep it warm, it wasn't insulated, and had no running water except for a hand-pumped cistern under the pantry. The toilet facility was a two-hole privy attached to the blacksmith's shop seventy-five feet from the house.

By the winter of 1964, four of us were living in that house. Glen Perry was a twenty-seven-year-old retired motorcycle racer, bulldozer operator, sometime welder and occasional motorcycle dealer. Zack Heenan was of Indian-Irish heritage, well-read and with numerous talents but no particular trade. Brian MacKenzie had considerable artistic talent. We all had overlapping interests. Wood-carving, metalworking, motorcycles, and antique guns were among them. All of us were working on this "big" commission I had landed and, as I had designed it, I was supposedly the boss. Of course there was a deadline and after a few weeks I could see that we needed to pick up the pace. Our working arrangement was loose, to say the least. I usually got up and organized some breakfast around nine, hoping the guys would be working by ten. Some days that worked, most days it didn't. As the winter wore on, the working day grew shorter and shorter and I grew more and more frustrated, trying to motivate the crew out of bed on frosty mornings. The deadline was coming closer and closer.

One morning after numerous calls I still could get none of them to stir. Finally in desperation I loaded blanks into a cap-and-ball Colt revolver I had acquired. In each room I fired a shot. It worked. The crew was up within five minutes. I had hot coffee ready and their breakfast on the table. There was rumbling and grumbling, but we were working by nine-thirty.

It worked so well that the next morning I loaded up the Colt and headed upstairs to give the boys their wake-up shot. Glen Perry's room was first. I lifted the latch and was about to open the door when I heard the distinct click of a gun being cocked. I swung the door open and jumped back as a shotgun blasted through the open door. I was down the stairs in a flash as a few more shots reverberated through the house. The three of them were up, clothed, and armed with shotguns and rifles, and they weren't using blanks. They weren't aiming to hit me, but it felt a bit like the gunfight at the O.K. Corral. That poor house to this day has quantities of lead embedded in the walls. The next morning I gave up on the Colt alarm clock.

However, I had started something, and the boys got out the old muzzle-loaders we found in the place and every now and then, like kids, we would have a big black-powder shoot-out — no lead, just wadding. The neighbours either pretended it wasn't happening or were too frightened to find out. This carried on sporadically for about a week until one day when I was in the two-holer outhouse during my regular morning business there came such a blast from beneath me that I was lifted clean off the seat and tumbled screaming, pants around my ankles, into the yard. Glen and Brian were rolling on the ground in laughter. They had loaded up the blunderbuss and thinking I was sitting on the farther hole, had poked the barrel quietly into the clean-out door. As it happened, I was sitting on the near hole, my vital parts only inches from the muzzle when they pulled the trigger. I had stinging powder burns and could not sit down for several days.

Our four-way partnership did not last much longer, and it was Zack and I who finished the Yorkdale commission — on time.

CHAPTER FOUR

Finding Paula

After a few years at the blacksmith's shop I was getting bored. It was 1966 and I was twenty-seven. Then some friends of mine decided to drive a Land Rover to Brazil and hired me on as driver-mechanic.

The only really great thing that came out of that trip was that in Mexico we met three blonde women at the artists' colony of San Miguel d'Allende. One, Paula Vockeroth, was six feet three and very beautiful. Just walking down the street, she would stop traffic. We soon were dating. I thought Paula might be twenty. She was actually seventeen. She was studying weaving and fabric arts while living with a family called Brigden, but she didn't like some of the others living there. I helped her with a grade 12 correspondence-school chemistry course and we became friends. Eventually I suggested, "Why don't you move in with me?"

"What would my parents say?" she replied.

"Tell them you're moving in with another family."

We lived together for the next few months and when

school was out, we took a train back to Toronto. Her parents picked us up, assuming that I had just travelled back to Canada with Paula.

Their family plan had been for her to go back to high school. We had other plans.

We told her father, who worked with the federal environmental instrument branch, "We're going to get married."

"What?"

He was an engineer and meteorologist. To him I was just a hippie artist.

It was during the debate about us getting married that Paula's father, maybe just thinking aloud, said, "This would probably be a good time to send Paula to Europe for a year." Obviously they thought a year away from me would bring her to her senses.

I said, "I think that's a good idea because I've always wanted to go to Europe myself." I guess they knew then that we were serious and weren't going to back off.

I can laugh about it now, but it was no laughing matter then. So we all compromised. Paula would go back to finish high school and we would agree to postpone the marriage for a year.

I got a job and Paula went back to school. I was broke, but found work at a welding shop to earn a bit of money. One day I saw a pile of scrap at the welding shop and thought, "I could make that into something," and six dollars later it was mine. My friend Gerry Fisher had just opened a refurbished restaurant called The Paddock in Greenwood. I lived in an apartment behind the restaurant. I told him, "I've got this scrap. I'll make an iron horse to go in front of your restaurant to attract a lot of attention. You supply the electric welder and I'll build the horse. All I want is the apartment in the back in trade, but if you ever sell the horse, I get half of whatever it sells for."

He said, "O.K., it's a deal."

So I built a life-size horse out of all this scrap and welded it together. Everyone quite liked this horse. I'm good at doing horses. It had an exhaust pipe for an asshole and a big crankshaft down the centre, all just made from auto parts. It attracted a lot of attention.

Paula and I were dating, of course, and when she got a summer job running the restaurant, I ran up a fairly sizeable bill there. We decided eventually we'd better sell the horse to pay my bill. However, we needed a little publicity to help sell it. Some friends of mine and I came up with an idea, sort of a switch on the old Trojan Horse story. I thought, "Let's put my horse in front of Toronto City Hall in the middle of the night, just for fun. That should attract attention." I devised a trailer to carry the one-ton horse and a base so we could get it unloaded at the other end. After the base was built, we drove the horse through Toronto in the middle of the night and unloaded it just before dawn right in front of City Hall in Nathan Phillips Square. When discovered, it naturally caused confusion. The City Hall maintenance crew didn't know what this creation was doing there. The whole bureaucracy was thrown into turmoil; they thought someone must have ordered it. But no one had and no one knew anything about it. People were asking, "Where's the security department? Nobody should be able to do this sort of thing!" The horse was featured on the front pages of all the Toronto newspapers and on the national television news and word got out that I had made it.

We brought the horse back to Greenwood. Then Gerry Fisher arranged for the horse to have a starring role at the annual Royal Winter Fair, a national agricultural and livestock event. Farm equipment people were trucking it around to display it in front of their booths. One day I got a call from Gerry: "You better come down here."

I found my horse. There was a crowd surrounding it, staring. Obviously it was a big hit. Prince Philip had opened the Royal Winter Fair that year. He went with a lot of directors to look at my horse, and commented, "This horse should be here permanently."

In the end, after much to-ing and fro-ing with the board of directors, they decided — in spite of Prince Philip — not to buy the horse. But the publicity sold other horses later.

Paula and I did wait a year, as we had promised. On June 27, 1968, we were married. That year was also a great year in other ways because the London Fair gave me a commission to do a horse for them, a $5,000 commission. Then a large hardware store gave me a $20,000 commission to do a mural in Ottawa. About $15,000 was cash and $5,000 was credit in their store. Now $5,000-plus was a lot of credit back then. We went out and bought a hundred-year-old schoolhouse for $10,000 at auction, a one-room building with the land it stood on. At least partly because of that horse, I went from being a pauper to owning a house and getting married. With the credit at the hardware store we furnished our schoolhouse home and equipped a studio.

Life went on serenely for a couple of years as I did more small murals for the hardware store and more animals for the London Fair. This gave us enough to keep going, since at the time we only needed about fifty dollars a month to live. Earlier that fall we had decided to take a trip to Mexico during the winter. Going south is cheaper than trying to heat a house in Canada! We sublet our house to a guitarist in a rock band and his family and converted an old Ford van as a camper, welding an Austin Mini top to the top of the van so I had a sort of conning tower. We put my Motocross bike into a trailer and headed out for California, intending to retrace our steps down the west coast and maybe spend the winter in San Miguel. We arrived in California short of money. I sold the motorcycle and got a job in Palm Springs repairing motorcycles for a couple of weeks. We then headed into Mexico and roamed down the west coast for most of the winter, heading home in February to find our house still occupied by the rock group and to discover they hadn't yet paid any rent, nor had they paid any bills. They were huddled around the electric oven in the kitchen to keep warm because the heating oil had been shut off. The result was that we returned not to the little grubstake of rent money we were expecting but a pile of debts for phone, electricity, and oil. After things were settled and we were back in our house, we applied for welfare.

I didn't want to go to my father for money and I didn't want to go to Paula's parents for money.

The secretary who ran the welfare office had been a secretary at Pickering High School when I was a student there. She remembered me because I was often sent to the office for disciplinary reasons.

When I told her I wanted to apply for welfare, she laughed, figuring it was a joke because I'd gotten a lot of publicity about my sculptures over the years. She thought I was rich; people equate publicity with fame and wealth. Paula and I were now on welfare.

Three months went by before history intervened, providing me with the impetus to plan and complete a project that I am still proud of. Let me go back a bit in time and tell you how it came about.

Highway 141 from Parry Sound to Bracebridge in north-central Ontario twists and turns through the rugged primal rock outcrops and crystal lakes of Ontario's Muskoka district. Paula and I were driving from farther north to our schoolhouse home a hundred miles to the south. Above us, Neil Armstrong and Buzz Aldrin had made their historic landing on the moon and were resting before stepping down to the moon's surface. When it became obvious we were not going to make it home in time to watch this monumental event on our own TV, we headed east to the village of Baysville, on the Lake of Bays, where on a friend's TV we saw the two earthlings step down on the Sea of Tranquility…, almost a quarter of a million miles away. No one really knows how many people were watching. My guess is that more were riveted by that moment than by any other event in human history. The depth of feeling was truly global, planetary and cosmic.

Before I was ten I knew man would reach the moon. A lot of people my age intuitively knew it. In one of my boyhood daydreams, I was the first man to land on the moon. It would be the adventure of my generation — an event greater than any in our history. As a high-school art project a couple of years before the Soviet launch of its unmanned Sputnik and a dozen years prior to Armstrong's "small step" to the surface of the moon, I painted a mural, my concept of the first moon landing. Now, years later, the reality of those men on the moon was overwhelming to me. In a sense, my daydream had happened — a great expansion of human consciousness brought home in those photos of earth rising behind the moon! There it was, our spherical jewel, our home, cruising through the cosmos, where billions of us lived, loved, fought, worked, and played, protected by earth's atmosphere. At Christmas my sister Alaine gave me a small plastic model of the lunar module.

Something clicked. I would build a full-scale replica! This image that is with us all is going to become an icon, I thought. This was my logic, to make it and sell it to a museum, a unique representation of the first thing that man had landed in on another planet. It would be a moonship on earth, a daily reminder of the beautiful sphere we inhabit, a monument to man's first extraterrestrial landing (and more practically, from my standpoint, a spare sorely needed room requiring no building permit). The idea turned to a passion. I had no money to launch the project. Being a sculptor does not generate much in the way of a cash flow at the best of times. Paula's eventual success with furs and fabrics in the fashion industry was still a dream, and my career was in a definite slump. But I had a friend who had made a lot of money in the school-bus business. I went to him, explained what I had in mind, then asked, "Will you fund this?"

He said, "No, I won't, but I'll introduce you to my

banker." So he took me into the Royal Bank and said, "My friend here, you've heard of him, great sculptor, has a project that he wants to do and he's going to make money on it in the long run but needs financing on it now. Will you lend him some money?"

"Sure, how much do you want?"

I didn't have any idea how much I needed. I said, "Oh, eight thousand dollars."

He said, "O.K., sign here."

That eight thousand dollars financed us through 1971. I teamed up with Robin Teeling, an old motorcycling friend and a tool-and-die maker currently unemployed, and we started building the Canadian moonship. We did have problems, mainly about design, but were damn near finished in the spring of 1972 when Paul and I got hit even harder. The federal government announced a plan to build a big airport in our area. along with thousands of acres and hundreds of other homeowners, our property was being expropriated in the name of the Queen!

We had been refused permission to build an addition to our old schoolhouse. Paula was now pregnant and we badly needed more room. We were torn between fighting the airport and taking the expropriation money so we could build elsewhere. It was certainly the only way we could replace our house. I was leaning toward the expropriation money when a newspaper reporter came to see me and changed my mind. By the time he had given me chapter and verse on various political deals involving airports in Montreal, New York and elsewhere, I was all fired up. I said, "What do you have to do with politicians, drop a bomb on them to get their attention, or what?"

He quoted that line and in the paper was a picture of me and Paula. I'm wearing recycled military clothes. My hair was still long. I had a beard — and I'm standing up there holding an old Lee Enfield. The headline said, 'LISHMAN READY TO GO TO WAR OVER AIRPORT!'

The next thing, my phone was ringing off the hook. I got dozens of calls from people who wanted to enlist in my war to stop this airport. I became one of the leaders in the airport fight.

From then on it was my war. My part was media events. Items: a mock funeral party at Queen's Park, the mourners all in black (Paula made the costumes) carrying wild flowers; a fake newspaper made to look like a government hand-out; and a lot of other creative anti-airport publicity. As of this writing, twenty-three years later, there is no airport and I believe there never will be!

During that midsummer of 1972 a number of my friends came to help me, not only with the airport battle but also to finish the lunar module. By then I had so little money left for the project that even our tools were scratch-built from scrounged material. My talented friend Zack made an ingenious shear out of a couple of discarded grader blades, while another friend, Bill Miller, fashioned a welder out of a furnace ignition transformer, some sheets of glass, and aluminum foil. With it he did all the aluminum welding. I found some industrial clamps in a junk sale and with them and some more discarded material, I created a sheet-metal brake. We forged on. I had researched the project as thoroughly as possible, even writing to Grumman, the original builder, and NASA for drawings. What I received was nothing you could call a working drawing, just some simple isometric sketches and limited photos. I don't think they grasped the reality of what I was attempting. Amazingly, it came together and sat gloriously complete by the time the last trip to the moon was staged.

During the airport battle I met Larry Miller, an aspiring filmmaker, who got excited about the lunar lander project, and together we raised a little more money and made a short film about building it. Once finished, it attracted all sorts of attention. Most people had seen pictures of the original NASA lander and most of them could not believe what they saw on our rural Ontario side road. They would slam on their brakes, back up, get out, and still not believing what they saw, invariably ask what it was. When I told them, they would usually say, "That's what I thought." Their curiosity then would really be wound up and they would ask what it was for and why I had built it. At that point, with our old school-house home under expropriation, I was not exactly sure myself, so I would attempt to make up a new reason every time. I told one tourist that the Planet Earth was going to blow up in a few years and I was building this spaceship to send my son to a new planet where his powers would be a thousand times greater than they are here. (Shades of Superman!) Another time I said I didn't know where it had come from, that I had awakened one morning and there it sat. On another occasion I told them that I was building a decoy to attract UFOs. We had great fun with it. Many people did.

One friend told me that long before he met me, on several occasions while driving friends back from his cottage late at night along a nearby highway he would yell, "Look, did you see that!" His dozing passengers would wake with a start, he would put on an agitated act about something weird-looking coming out of the sky, then turn up the side road and drive up to the lunar lander, which I kept rigged with lights and scorched grass as if it had just landed. People would knock on our front door in the middle of the night and excitedly tell me that a spaceship had landed in my front yard. I would say, "Yeah, I know. They are from Uranus, they lost their way and wanted to use my phone." Then there were the drunks or people high on something else who would wander in and come out with original observations like, "Wow, outa sight" and "Far out, man."

I had at that time installed a special table in the schoolhouse that retracted up into the thirteen-foot-high ceiling. It was quite practical, for it converted the one room of the school from living room to dining room at the flip of the switch. It was also great fun, for we could set the table and put it up out of the way, then when the guests began to get fidgety and wonder whether or not they had really been invited for dinner, flip the switch and down would come the table laden with food kept hot with lamps in the ceiling. The lights for the table, which were in spherical reflectors, were hung on counterweights and also retracted with the table. For some reason they would stay hung up for a few seconds after the table had come down, then gracefully bloop down individually, coming to a gradual stop at just the perfect height above the table.

One Saturday night, a friend and I for some reason had both donned weird hats out of my hat collection. We were sitting at the table. He had never seen it retract, so I put it up and of course the lights went with it, leaving us in the dark. In a moment I flipped the switch and just as the table began to descend, some rather stoned-looking people sauntered in the door, attracted by the illuminated module nearby. They stood in the doorway watching this surreal diorama of the table delevitating out of the ceiling between these two weirdly attired, bearded characters. They looked at each other in absolute disbelief and without so much as a word stumbled over each other as they rushed out the door.

We had a visit from CTV's news anchorman, Harvey Kirk, who stopped by like any other tourist. He had covered the moon landings for CTV from Houston and was intrigued by the lunar lander. We kept in touch and Harvey became the narrator for the film we made about creating the lander.

The filmmaker had great plans. He organized meetings in New York and we flew down and showed the film to the three major networks. CBS bought it and in Canada, CTV bought it for the public-affairs program "W5". We were elated. It was an eight-minute film, but both CTV and CBS cut it so that it was unrecognizable as the film we had made.

CBS aired it nationally during the Apollo 17 touchdown, and a month or so later we received a cheque from them for one hundred dollars. CTV paid fifteen times better. However, our out-of-pocket cost on the production of the film was almost two thousand dollars. It was obvious that neither Larry nor I had been anywhere near the Harvard Business School and our filmmaking ground to a halt.

The following year my family moved to Purple Hill, twenty miles or so east of the old schoolhouse, and now the appropriate thing seemed to be to sell the module. But how? We tried a mailing campaign to all the science and aviation museums in Canada and the United States. At first we got a heartening response. The most enthusiastic came from the curator of the museum in Neil Armstrong's hometown, Wapakoneta, Ohio.

Again we received a lesson in business. Museums, it seems, rarely buy items. Individuals and corporations donate artifacts to museums and use the donation as a tax write-off. Just what every starving Canadian artist needs!

I moved the lunar lander to our new home and left it disassembled because I didn't want a repeat of the tourist influx. It became my white elephant, my personal albatross, and as the years went by, many people ridiculed me over the project as it lay corroding in pieces outside my new shop.

Twelve years after beginning the lunar project, I got a phone call from an ex-World War I pilot named Clarence Page of Oklahoma City. He, along with ex-astronaut Tom Stafford, had created the Oklahoma Aviation and Space Hall of Fame. They were among those who had previously contacted me about donating our lunar lander to their museum. He wanted to know if I still had the module and then told me that he was sending some Japanese people to see me and I could make whatever deal I wanted. This was to be another lesson in the world of business.

A few days later, Mr. Tanaka flew in and I met him. I had scraped the weeds off the lunar module's parts and propped a few pieces up. There was a derelict air about it, but still it was unmistakably the lunar module. On the way from the airport, Tanaka asked me how I had managed to get a lunar lander from Grumman. I assured him it was Canadian made, yet identical in size and shape. Like most Japanese businessmen I have met since, he kept his cards close to his chest. He didn't say much, simply asked for the price and left. Two weeks later he called and told me his company would like to purchase the module. He wanted me to meet him in Washington at the Smithsonian Institution in two days. I was there, and in front of the only remaining real lunar module, we signed a deal.

I was to refurnish the module and have it ready for them in six weeks. It was to be part of the Great Space Shuttle Exhibition in Tokyo, in honour of the twenty-fifth anniversary of NASA. It would be the main artifact for the "yesterday" part of the theme, which was "Space, Yesterday, Today, and Tomorrow".

"Also, Mr. Lishman, we would like you to attend opening and be honoured guest. We will fly you and wife first-class to Tokyo and you will stay at Palace Hotel, part of old Imperial Palace. O.K.?"

O.K.! No more white elephant. No more albatross! It had turned to gold! I was so elated about selling it after twelve years that I took off for New York, where Paula was conducting some business. We bought a bottle of the best champagne and celebrated.

We really had to hustle for the next six weeks, however. With the help of a couple of assistants we completely restored the lander, even making extra pieces that I had discovered while viewing the one at the Smithsonian. I delivered the lunar module on time and we flew to Japan first-class. The exhibition opening was full of pomp and ceremony, attended by the U.S. ambassador and Japanese royalty. Paula and I had a front-row-centre seat and were introduced as members of the NASA lunar module exhibition team — we did nothing to correct the error. When we were introduced to the U.S. ambassador, he asked, "Where are you-all from?" I told him Blackstock, Ontario. He just looked puzzled and walked away.

But there is a little more.

What I didn't know was that the gentleman from Oklahoma knew the Japanese were desperate to find a lunar module and had worked out his own deal. I think he had been to business school. At great expense, the Japanese had acquired the original mock-up of the space shuttle Pathfinder, dismantled it, shipped it to Tokyo, built a special building to display it, and had searched the world for a lunar module. All the originals save the one at the Smithsonian are still on the moon. I had the only other one on earth! Clarence Page and Tom Stafford had made a deal with the Japanese that if they told them where I was, then when the Great Space Shuttle Exhibition was over the lunar module would be returned to rest permanently at the Oklahoma Aviation and Space Hall of Fame.

Our daughter, Carmen, was conceived in New York on the evening Paula and I celebrated the sale of the lunar module, twelve years after its conception. When I look at Carmen today I smile and think about a day when she is a little older when I will take her to Oklahoma City and show her my replica of the first spaceship that man used to land on another planet — perhaps on her twelfth birthday.

CHAPTER FIVE

Ultralight Romance

I was beginning to get established as a sculptor, but I had not forgotten about flying. The early seventies gave rise to the first crude hang gliders. Now that looked closer to flying like the birds! I met Mike Robertson, a neighbour who had become a local celebrity performing outrageous feats while hanging from a flat kite towed behind a boat on Lake Ontario. It was Michael who reawakened my desire to fly, for he was full of enthusiasm for this new kind of flight.

I was amazed to learn that the origins of hang gliding dated to well before the Wright brothers. A German, Otto Lilienthal, after twenty years of studying bird flight and experimenting with models, made the first flight with a man-carrying glider. Lilienthal ran with the craft on his shoulder and when it lifted him he hung by his armpits, thus inventing the term "hang glider". Lilienthal even created a hill to carry out his flying experiments, but hang gliding suffered a setback when he was killed in 1896 in the crash of one of his pioneering craft.

John J. Montgomery, an American, is really credited with the beginnings of modern aviation. Early air pioneers had generally used flat wing surfaces, but Montgomery studied bird wing sections and discovered that the cross-section of a soaring bird wing is a parabola. Thus he had the breakthrough of discovering the curved wing, making conquest of the air possible. In 1883, Montgomery took off from the rim of the Otay Mesa south of San Diego and made a flight of 603 feet. Twenty years later the Wright brothers made their first powered flight. From that time on, experiments in slow, quiet flight seem to have been abandoned in favour of the quest for speed, altitude, and distance. Civil aviation evolved through its romantic era into what many came to regard as practical, overregulated boredom. Aircraft had developed way beyond the wonderful flight envelopes of birds. In 1949, Francis M. Rogallo patented a simple delta-shaped flying wing made of tubes and fabric. His idea was that it would be a replacement in many ways for the parachute, but would have much better directional control. The Rogallo wing languished through many attempts to make it usable by NASA and the military, but it never became a practical replacement to the parachute.

In the late sixties Bill Moyes, an Australian, adapted the Rogallo idea as a flexible man-carrying flying wing. His partner, Bill Bennet, introduced it to the United States on July 4, 1969. Bennet was towed up the Hudson River by a boat and then flew the Rogallo delta glider over the Statue of Liberty.

It did not take long for the romance of the air to be rekindled and soon there were all kinds of backyard, would-be aviators experimenting with the ultra-simple Rogallo-concept wing. The Wills brothers from California made wings from bamboo and polyethylene film. Dave Kilbourne, another Californian, was credited with perfecting the foot-launch of these early flimsy machines.

In eastern Canada, Michael Robertson was the pioneer of hang gliding. An athlete and a showman, Mike had graduated from flat kits to towed Rogallo-type gliders as developed by Moyes and Bennet. He started making his own Rogallo hang-gliders that could be foot-launched from a hill. Michael was the flying star of the water-ski show at the annual Canadian National Exhibition, and the day after the show closed in September 1972, I strapped on a pair of water skis, was harnessed to one of his Rogallo delta gliders, and was pulled up into a very wobbly flight at the Toronto waterfront. It lasted only for a few minutes until I went too far, ending the flight with a splash reminiscent of the pond landings of Whitey the goose. But it was my first real flight. Once again, I became obsessed by the idea of flying.

At Purple Hill, named after the viper's bugloss that adorns the hillside, I acquired one of Michael's backyard-built Rogallo gliders. Under his guidance, I proceeded to run off the knoll behind the house until I got thoroughly bruised and somewhat proficient. In later years Michael would refer to me as his worst student.

The early standard Rogallo hang glider was a tricky thing to fly. It had a glide ratio slightly better than a sheet of plywood, but less stability. If you got the nose off-wind slightly on take-off, you were back into the hill in a flash, but after many tries I did achieve a brief few seconds of real flying. But I soon realized, that without a handy mountain to fly from, gliders would need a source of power to get them to a decent altitude, so at least you would have some time to look around.

I read about the Icarus, a biplane hang-glider out of California that had flown off the ocean cliffs for more than five hours. Now that sounded more like bird flight. By the time I had saved the money and bought the kit, the Icarus had evolved into the Easy Riser hang-glider. That summer, 1976, a friend invited me to join him in his Cessna 150 for my first pilgrimage to Oshkosh, Wisconsin, the absolute world capital of experimental aircraft, where I had the luck to witness something truly inspirational. A man named John Moodie had bolted a go-cart engine and propeller to the back of his Easy Riser and there he was, running off flat ground and cruising into the sky just like a bird. It was a historic, awe-inspiring foot-launched flight.

The image of this man unfolding a pair of wings, bolting on an engine, then picking it up and running off into the sky had all those watching totally amazed. It was the realization of mankind's age-old dream of bird flight. It truly was a milestone in the "reinvention" of flight. I use the word reinvention, for that event heralded the serious beginnings of the ultralight aircraft movement. A new breed of aviator had returned to the time of Lilienthal and Montgomery, to pick up the missing part of manned flight that had been bypassed since the time of the Wright brothers. A new romance with the air had begun. By 1978, in an alfalfa field owned by our neighbour, Harvey Graham, I was able to duplicate John Moodie's flight.

In my first year of Easy Riser flying I met Jack Weber, who had built one. I cannot count the number of times in those early days when we would join forces in Harvey's field and fly, or try to. It took several bent wings and much frustration to successfully add an engine to a hang-glider. In those early powered hang-glider days, the real thrill (once you overcame the fear factor) was to run off the ground and take off like a bird.

On a typical take-off the wings sit nose down into the wind, tilted forward and propped up by the propeller guard/skid. If the engine was feeling right, after a few pulls it started and idled erratically. Getting into and then settling the aircraft up on your back into position was the most difficult part. With the engine running, you had to climb in through the cramped front down tubes, then squat down between the parallel control bars, grasping the bars with one hand forward and one back and raising the nose of the plane. If there was any breeze, it helped the lift. Then once everything was balanced and the aircraft pointed absolutely dead to the wind, you'd grip the kill switch between your teeth. (It was called the spit switch because with both hands on the control bars the only way you could kill the engine was to spit out the switch.) It was a safety factor, ensuring that the propeller would not make hamburger of you if you stumbled on take-off.

The hard part was to lift the entire aircraft up and get it balanced on the small of your back while keeping dead to the wind. That was no mean feat, considering that the craft weighed more than a hundred pounds and had near-ly a thirty-foot wingspan. If the nose got too high and the wind was a little brisk, it tried to throw you over backwards. Similarly, if the nose got too low, it added immensely to the weight.

Once in position, feet apart and bracing yourself against the backrest, you had to hold it at the perfect angle of attack. Then quickly moving your left hand and letting that side sag a little, you hit full throttle. The engine revs to a scream, the propeller turning a few thousand RPM six inches from your butt. In a moment the thrust is too much to hold back. It is either go forward or fall flat on your face. One step, two, three — with each step the load gets lighter, your strides become easier. Within a hundred feet the wings are flying free under your armpits. Another few feet and you feel yourself lifted. The wings are beginning to take you away faster and lighter, faster and lighter. Your legs move with ease as the weight is lifted from them. You are running faster than you would have ever thought possible, great long wonderful strides, feet just brushing the ground and at last totally free of the ground. Your legs, not believing it, are still flailing open air. You are free, truly airborne. The field drops away beneath you and a great vista opens up. I have never experienced such a contrast in extremes — at one moment fear-weakened, overburdened, and awkward, clamped down by gravity and entangled in a banshee-screaming, cumbersome contraption, then within seconds becoming lighter and lighter, freer and freer. Even the noise seems to drop away with the gravity in the most graceful manner.

That is, if everything goes right.

Many times things went quite wrong. There was the time — my sixth powered flight — at about a hundred feet, when the right wing stalled and dropped away.

Pointed straight down, I spit the kill switch. Miraculously the Riser pulled out and I was skimming the ground downwind at thirty miles an hour with a ten-mile-an-hour wind. A downwind landing then would be a disaster, for the full weight of the aircraft would be on me while I would be attempting to run on foot at twenty-something miles an hour. I attempted to bring it around to the wind. The right wingtip caught the ground and did two marvellous big slewing cartwheels into a jumble of tangled aluminum.

It was almost a year before my next attempt and further mishaps, like the time when the tip of my heel connected with the tip of the propeller on the final paces of take-off. I had devised a safety guard to prevent such occurrences, but in the long final strides of take-off the heel of my shoe rotated around the guard just enough to clip the tip of the prop. The prop exploded and took out most of the right lower wing. I spit the switch and rolled into a ball of bent aluminum and fabric. I survived with only minor bruises.

In the years that followed I built and rebuilt, tried different engines, carved prop after prop, and bought and made reduction drives. In short, I lived the romance and evolution of this new form of aviation and my Easy Riser became a refined little plane with a three-cylinder, electric-start radial engine, an elevator, one-stick control, and steerable tricycle gear. With meagre funds I built a hangar and put in a six-hundred-foot airstrip. It was the only way to enjoy flying that little craft, for its wing loading was so minimal — less than two pounds per square foot (a modern airliner is ten to twenty times that) — that any turbulence was keenly felt. And I only wanted to fly when it was enjoyable. So the aircraft stood ready. When the weather was right, I could be airborne in short order.

The best scenario was to rise at daybreak and take a cruise before the regular day began and then close the day watching the sunset from my own private mobile mountaintop. The Riser cruised nicely at thirty miles an hour. Technically it looks flimsy as a butterfly, yet it is extremely strong. Being a biplane, it has the strength of a box and also the drag, which prevented me from flying at speeds beyond its limit. For several years I cruised the local countryside, observing the changes in seasons and occasionally watching flights of birds from my birdlike vantage. Many mornings before breakfast or on clear evenings before sunset, I would wander off through the air currents like a lone hawk, turning lazy circles a few hundred feet in the air, absorbing the countryside, living in that sea of air before the ever-changing terrain. In early spring when the grass has just turned green and the hardwood forests are transparent without their leaves, it's possible to see how the land was formed. I discovered many mansions tucked away in wooded lots and saw the personality of each farmer written in the patterns of the fields. On occasion I would come on a flight of gulls or would see a blue heron and fly behind them at a distance. Several times I was able to catch a thermal with a hawk and spiral up a few hundred feet, for they have no fear of the aircraft. I always felt that they just considered my craft another bird. Numerous times I was within feet of them. And always the lure of flying with the birds was there. Even though I knew I wasn't really making much headway, the dream wouldn't go away.

True flying with the birds began for me on one of those autumn dawns when pools of mist lie in the hollows, the air does not move, and the long shadows of the fence rows are drawn across the stubble fields by the rising sun. The farm woodlots of southern Ontario were beginning to show splotches of golds and reds among their summer greens. I had flown east from my airstrip at Purple Hill in the Easy Rider — absorbed in the clear crisp beauty of the day. To the north, Lake Scugog sat mirroring wisps of vapour, the rising ghosts of yesterday's warmth. After about twenty minutes a long, lazy semicircle pointed me homeward at about five hundred feet. As I approached the Swain farm I noticed a stubble field almost black with birds. Throttling back, I started dropping down for a closer look. The birds sensed my approach and rose as one, like a blanket of ducks, thousands of ducks, and in a moment I was in their midst. I powered up to climb out of the flock, but they were climbing too, our airspeed matched. They made no attempt to dive away from me: by default, they had accepted me into their number, save the two that were directly in front that kept glancing back nervously, maybe worried that this might be some new predator that would devour them in mid-air. When it became obvious to them that I was not a threat, they just winged on. The air was full of ducks — in front, below, above, behind, some in ragged chevrons, some in amorphous clusters, wings flashing in the early light. The thrill was indescribable. Caught in their mass spirit I winged along, just another bird in this autumnal squadron headed for the marshes to the northwest, my attention focused on a duck off to my left holding perfect formation about four feet off my wingtip, just as if I had always been his wing mate — and for a moment I felt I had.

I drank in this wonder, the experience one of those that transcends time. Below and to the south, through a myriad of ducks, I glimpsed my house and airstrip slipping by, as if they were a child's model seen on the bottom of some great river of ducks.

As we went on, there was a flow within the flights, a movement of groups smoothly shifting their positions as if under the baton of some unseen conductor. Then the open fields gave way to the dark green of the forest and there, with this great massing of wings, I was in a time before man had surrounded the planet, filled with a primeval feel dating from eons, ages, before freeways and television.

I became aware that we were descending. I too had tuned in to the flight conductor.

The reality of my mechanical bird wings interceded, and the fear of my engine quitting over the woods brought me out of this marvellous trance. Reluctantly I pushed the throttle forward and climbed up, through, and out of the mass of wings. They flowed aside, letting me through. I winged over to the left full of what at that instant I didn't have a word for, and now can only describe as glee, I swept around to head home, watching the sparkling of wings receding lower and lower into the dark scrub of marsh that had not yet received the sun.

In a few moments I was rolling down my bumpy airstrip, shaken back to ground reality, feeling that I had for a moment experienced something of great significance. I could not get this joyful feeling to leave. I kept my flying machine at the ready, craving to spend more time with these creatures in their element, and experience the world from their vantage again.

It was not to happen that year. Once, returning from a sunset flight, I did spy a flock of about thirty Canada geese, higher and a little off to the south from me. I swung around and made an attempt to join up. It was not their wish. Even with my machine at full throttle they gradually outclimbed me into the twilight.

The snows came. I put the Easy Riser away in my makeshift hangar (capital cost, twelve dollars) and busied myself with the major sculpture commission I was working on for Expo '86 in Vancouver. Then, later that winter, disaster struck, neither the first nor the last in my long affair with ultralight aircraft and geese. The hangar collapsed under a huge snow load and completely crushed the Riser, putting that part of my life back to square one.

So where should I go from there? Get another aircraft, obviously! I immediately started thinking about getting a Lazair, a second-generation ultralight designed by Dale Kramer and the late Peter Corley when they were little more than teenagers. For a few years it was the darling of the ultralight crown, the aircraft to have. In its original form it had two six-horse-power converted chain-saw engines and one stick that controlled everything. The Lazair had long, gracefully tapered wings with a little upsweep at the tips, and an inverted-V tail. It was and still is a beauty in the air, both visually and to fly. It also afforded wonderful visibility. I must admit I lusted after one for years. On the demise of the Riser, I found a good used Lazair for sale. I paid half down to clinch the deal, and one fine day in April 1986 I went to take delivery of it a few miles from my airstrip. The man who was selling it to me flew it over to a huge sod field a few miles from our home, a good level area for me to make my maiden flight. Like most ultralights of the time, it was not a two-seater. The usual procedure was for the instructor to stay on the ground, giving lots of advice and watching the student learn.

The day was perfect, without a wisp of wind. I thought flying it would be a piece of cake. After all, I had had several hundred hours in the Riser by then and assumed I would not have a major problem converting to the Lazair.

Wrong assumption! This was a Mark II Lazair with rudder pedals, an overhead stick, twin nine-horse engines, and narrow landing gear. The Easy Riser was a single pusher-prop, with a single conventional stick and wide, easily steerable nose gear.

Terry, my instructor, instilled great confidence in me. He flew the craft up the field at about three feet of altitude as steady as a rock — yellow prop spinner, yellow wheel fairings, a flying dream. He landed and walked around the aircraft, instructing me on taxiing procedures and so on. I think he told me what to do. I think I just did not listen well enough.

I got in, buckled up, pulled the engines to life, and started taxiing down the field. Then I suddenly found myself at a heading ninety degrees off my planned path. I got the craft turned around and suddenly was going in the wrong direction again. Overanxious, I gave it too much throttle. In a second I was airborne! I assumed it would be easy once I got it up in the air, so I climbed, my legs vibrating with nervous fear. I tried a turn. I moved the stick and the Lazair banked the opposite way to what I had expected. Confused, I pushed the rudder pedal. Trying to think it out, I reached down and pulled back on the throttle. No! No! With more throttle now, I was going in a tighter and tighter circle. Desperately I moved the stick the opposite way. I was panicking, but as in all panic I did not recognize it. A huge tree loomed in front.

Instinctively I pulled back on the stick to clear the tree and screamed as I stalled into the trees on the other side.

Crash! One engine was still running. I was sitting straight and level. The left wing was hugging the trunk of a tall poplar. The right wing sat on top of a smaller tree. I shut off the engine and sat there not moving. For ten minutes I sat there, thirty-five feet in the air, not totally absorbed in the beauty of the fresh buds on the branches and the sweet spring song of redwinged blackbirds.

Too soon, Terry came crashing through the underbrush, yelling my name.

I called out to him, "I'm all right."

But the fact was I did not want to move. I just wanted to sit there and enjoy the spring from my birdlike perch. Also, I had absolutely no energy. None. Some other people came through the underbrush. I did not want to see them either. They were carrying a ladder and poked it toward me. The ladder was ten feet short. I jiggled the aircraft to no avail. It was there solidly, a broken branch of the poplar hooked right through the leading edge of the left wing.

I looked down. The very top spike of a cedar was at my hand. I still did not want to move. The people on the ground were getting impatient and I was feeling more than a little embarrassed. I had only paid Terry half the money for the plane, so I made light of the situation and called to him that I had only smashed up my half and his half was still O.K. One of the people who brought the ladder called up, "We heard you scream over the engines as you went in." I don't remember that! No, that could not have been me!

Finally I made my Jell-O-like body start to work and attempted to clamber out of the seat and over to the poplar. I grabbed a branch. My hands had no grip. I had to concentrate just to hang on. It was probably the scariest thing I had ever attempted in my life. I finally got to the trunk of the poplar. It was straight as a telephone pole down to the top of the ladder, with no branches. Somehow I managed to hug it enough to lower myself to the ladder and get to the ground. Shaky, shaky. Terry drove me home. I had a stiff Scotch and went to bed for a few hours. I awoke at 3:00 A.M. thinking about how I was going to retrieve the bent Lazair from the treetops in the middle of a forest. Then I remembered a pair of old lineman's climbing spurs, the type used for climbing power poles. Flashlight in hand, in the middle of the night in the back of the shop, I rummaged through the considerable collection of odds and ends that I never throw out, and by 7:00 A.M. I was back in the forest with two helpers. My climbing gear consisted of a block and tackle and the old swing seat from my Easy Riser. Within half an hour we had the Lazair lowered to the ground and carefully disassembled. A piece at a time we carried it out of the bush and trucked it home.

I found another Lazair, brand new, but the engines had been stolen from it. The owner was desperate for money. I put together all the available funds I could muster and bought the Lazair at a bargain price. Within two weeks I had fitted the wings of the new riser on the fuselage of the crashed one, installed the old engines, and with the help of instruction from Jack Weber by two-way radio I was airborne, this time very carefully.

But I still was not much closer to my original dream of flying with the birds, actually flying among them.

CHAPTER SIX

Bill Carrick and Konrad Lorenz

Almost by chance, I met a man who was to help me a lot in the next stages of my attempts to fly with the birds. I read a story in a local paper about Bill Carrick, who had trained a flock of geese to fly behind his motor boat.

I sought him out — a soft-spoken, slight man — and asked him if he thought it possible to get geese to fly with an aircraft. Unknown to me, he had had the same idea, the same desire. He was following the experiments of Konrad Lorenz to, as he put it, "condition" Canada geese to fly with a motor boat for film work. Carrick cooperated willingly and taught me about "imprinting". This was a theory researched by Lorenz, who shared a Nobel prize in 1973 for his pioneering work in animal behaviour. The Austrian naturalist, whose own home was often filled with animals he was raising and observing, showed that goslings, soon after they broke out of their eggshells, began to attach irreversibly to the first moving object they encountered. Normally, this was a mother goose, but when hatched artificially, the geese could as well imprint on a human — or, maybe, on an ultralight aircraft. Sigmund Freud had independently arrived at a similar theory of some humans' obsessive attachment and response to objects, which he called "fixation".

Lorenz found that imprinting was a complex process, which took substantial reinforcement to become permanent, and was as dependent on sounds within a certain range as it was on sight. He described its first stages as the mother goose responding with cackles to the "lost piping" sound of the babies: "…When you first feel lonely, utter your lost piping, then look for somebody who moves and says, gang, gang, gang; and never, never forget who that is, because it is your mother." The famous animal behaviourist in 1933 spent several days in his laboratory quacking incessantly in mother's approximate tones to a just-hatched flock of mallard ducklings to show the role this played in imprinting. Fortunately, it was during a holiday, and other people in the building were gone, Lorenz wrote.

More recent research on imprinting has shown that it is virtually universal, but nearly as varied and complicated as the number of varieties of birds. Some, like the curlew, seem averse to attaching to anything but the real thing, a mother curlew. With cranes, it seems to occur over a much longer time period, months compared to weeks in geese. Experiments underway by the U.S. government at the Patuxent Wildlife Research Center in Laurel, Maryland, are attempting to imprint cranes to people in unique, bright-coloured clothing. The further plan is for these "mother cranes" to induce the birds to freely board a truck that will transport them along migration routes, periodically letting them out to fly and learn the route.

Birds that never imprint to anything, as may happen when hatched in an incubator, are likely to be behaviourally dysfunctional — not unlike human babies denied a normal measure of contact with a parent in early developmental stages. From everything that was known, the geese seemed an ideal species to experiment with; and if geese could be imprinted to follow a plane, there was reason to think it could be done with swans and even cranes.

Carrick would remove eggs from the nests of several birds in his domain and incubate them, arranging that he would be the trusted "mother" when they hatched. He knew that removing eggs would be followed by the bird laying more eggs, a natural system known as double-clutching.

When I first went to visit Bill at his eleven rented acres and nine ponds on the outskirts of Toronto in April 1986, I was greeted by a cacophony of goose honks, owl hoots, hawk screeches, and a number of other noises that I could not identify.

Bill Carrick's life had been devoted to the study of animal behaviour. He supported himself over the years through photography and film work, doing everything from supplying animals for nature films to making several of his own films. Anyone with an injured bird or animal would bring it to Bill and he would provide for its convalescence. As a result, his place was a funky jungle of overgrown ponds, makeshift cages and buildings. A small barn served as his headquarters, an organized jumble of ancient incubators, film equipment, and pens containing for the most part several families of beavers, as well as all shapes and sizes of cages made from a variety of scrounged materials and containing a broad variety of animals and birds — great horned owls, raccoons, foxes, red-tailed hawks — and large netted-over ponds containing waterfowl by the dozens. To add to this there were a multitude of wild geese and ducks free-ranging throughout the place.

Also on his property was a large, windowless, monolithic box of a steel building. It was a film studio. It had been built a few years previously for a dramatic film on the flight of a goose. The film had never come to fruition although megabucks had been spent on its preparation. The building housed a piece of equipment two stories high and powered by two V-8 truck engines, a wind tunnel in which a bird could fly at the same time as being filmed. A wheel twenty-two feet in diameter was designed to rotate around the flying area, carrying a movie camera on one side and on the opposite side a mounted screen. Any prefilmed background could be projected on the screen, and the bird flying stationary in the blast of the wind tunnel would appear to be flying over whatever was on the screen. The whole rig had been tested, but with the demise of the film it now just took up space. It was one of Carrick's toys that he seldom used but loved to show off.

When I told him I was in the early stages of trying to get geese to follow an aircraft, he was interested and laid out a plan.

The first step would be imprinting. Carrick hatched out a flock of goslings in the spring of 1986 and moved them to my property, where they would adopt our family — me, my wife, Paula, and our sons, Aaron and Geordie, and daughter, Carmen. We all worked with the goslings daily. Our plan was to first imprint them on a motorcycle and then try to convert them to follow an ultralight aircraft.

The young birds soon were firmly attached to me. At first we kept them under a heat lamp in the garage, for there can be very cold nights in early May. They grew exceptionally fast and after a week I moved them to a netted-over pen near the runway. From day one they were full of exploratory enthusiasm. They scurried after Carmen and me at every move; when we stopped, they investigated all that surrounded them. Their little bills, not quite rid of the spur that was used to open their eggshell, were pulling and pecking at everything. The taste and texture must tell goslings whether to tear off the morsel and devour it or to pass on to the next leaf, bud or blossom. They are compelled to tug and pull at everything throughout their lives. Later in their development they strip the seeds from grass stems, running their beaks up the stem in long stroking movements of their necks. Nothing goes untested in their passage — sand, grass, weeds, shoelaces, hair, bits of paper, insects. They even pick mosquitoes out of the air. These flurries of oral activity subside once the local area has been pecked at. Then they sit and tuck their heads back, sometimes putting their bill under a wing to rest, taking little catnaps. They do this as a group, as if one of them says O.K. now is nap time or now is explore time or now is preening time.

There is a constant verbal communication going on among them; it is not really cheeping, more a subtle three-syllable whistle/warble, but if one gosling gets separated from the flock it is cheep-cheep-cheeep-cheeeeep-

cheeeeep incessantly until it once again is reunited with the group.

When the young goslings were first introduced to a small pond, a new bit of genetic programming took over. At first they were cautious. They would hesitantly step into the puddle and stick their beaks into the shallows, testing and vibrating their beaks, then they'd get the idea to jump in and automatically swim, floating like neat little boats. Soon they were all in, paddling about, and then something would just snap in one of them and it would dive and swim all around under water, going like some crazed water bug. When it popped to the surface, it inevitably would spook the rest, causing them to dive. The action became a chain reaction, all of them diving, cavorting, splashing, and propelling themselves underwater like torpedoes, bumping into each other, creating flurries of splashing water. Then they started upending. As they investigated the growth on the bottom of the pond, their tails would stick vertically out of the water like some weird plant. Once out of the water, they would automatically go into preening mode as a group.

Inducing them to fly with my motorcycle was relatively easy. I spent an hour each morning and night on the airstrip having them follow while I drove up and down. By mid-July they flew with me daily. They became attuned to the sound of the motorcycle and would match its speed perfectly. When I revved it up to take off, they would be airborne in a flash and would cruise along so close that the leader's wingtip would be only a few inches from my cheek. In some instances, one goose would cruise just about three or four inches above my head. From my view, its head and neck protruding forward were like the peak on a strange hat.

The routine was simple. At about six-thirty each morning I would arrive at the pen on the motorcycle, a 200-cc four-stroke Honda Enduro model. I would position the machine, idling, ready to roll, about five feet in front of the pen. The geese would be eager to go, so when I opened the gate, I had to mount quickly and ride off or they would get way ahead of me.

The opening to the gate was off the main runway and about a hundred feet back. I had to make a sweeping turn onto the runway and accelerate to whatever cruising speed we had decided upon. If there was a tailwind, I would have to go a lot faster than if the wind was coming at us down the runway. Sometimes in the beginning of the flight the geese would all be in close proximity. I was constantly amazed at their agility in the air. They would bunch up in the turn, and although they came exceptionally close, they never once bumped each other or me. At the end of the runway, in the initial stages, I would brake to a halt and they would all land around me. When I decelerated they followed suit. I would turn the bike around and after a short respite rev it up a couple of times, which signalled them for the return flight.

Take-off seemed to be a democratic affair. If a majority of the group did not want to fly, the few that were airborne already, as soon as they were aware that the rest were not coming, would abort their take-off and a great cackling would ensue. Sometimes it would take several attempts. We would do a return flight back down the runway and then I'd either herd them or lead them back into their pen, depending on how rushed I was for the day. The whole routine was then repeated with a second flock that was housed in an adjacent pen. The second flock never flew as predictably as the first, for I probably did not spend as much time with them as the first group. In one of their initial flights, they took off away from the runway and only about three of the seven returned immediately. The rest got too tired and landed willy-nilly in the long grass. There were some anxious moments as we searched for them. Eventually they were rounded up and happily returned to the pen.

As the geese developed, they wanted to fly farther than the run up and down the runway, and on the occasions when the hayfield next to the runway was mowed I would make long, sweeping, curving runs over it with them. At the end of the field I could never make that long turn and had to slow down and turn sharply. They would sweep around and catch up as I accelerated back. Progress, but we were still working with the motorcycle and my aim was to get them to follow an aircraft.

About the end of July, I brought out my best ultralight, at the time the twin-engined Lazair (which by then I knew how to fly), and attempted to get them to follow it — on the ground. The giant wing and the strange noise completely spooked them. No way would they follow it. So we tried a new angle. My older son, Aaron, would ride the motorcycle and I would taxi the aircraft, hoping they would catch on that way. We made numerous attempts. It was difficult to get the three elements — geese, motorcycle, and aircraft — timed right; either the geese would get away ahead of us or they would be in sync with the motorcycle and the aircraft would be behind. Eventually we did get our timing right and they would take off, but instead of following the aircraft they would always return to the sound of the motorcycle.

The final attempt at this occurred in mid-August. We got up — aircraft, motorcycle, and geese. We started down the runway, aircraft and motorcycle in close proximity, the bike riding off my right wingtip. Looking back, I could see the geese flocked about the bike, but I did not see one goose fly up over my left wing until I heard a "brap blap" as it went through my starboard propeller. I was only about ten feet in the air when the disaster happened. The prop exploded but there was no problem setting the aircraft back down. The sad thing was that the goose was decapitated. That was a black day. We all sat down to rethink our plan.

I knew that we would have to make major modifications to the aircraft if it was to be safe with a dozen geese flying about. By the time I had designed, built, and tested cages for the props, it was too late in the season. We made several more attempts in the fall to get the geese to fly with the Lazair, but success was minimal.

About the end of October I did have one memorable flight. I took off and had Aaron release the geese. After I was airborne they did climb up and we flew for about ten miles somewhat together on a long circuit. However, I was more or less keeping up with them. When I flew out front they never really followed me. We never were closer than one hundred feet at any time.

We did try some interesting experiments with the motorcycle, however, I bought a new bike in mid-August and started using it. The geese would not follow it at first. It looked the same but the sound was different. We had to bring back the old bike and Aaron and I rode together with them until they accepted the new bike. We trucked them to the local fairgrounds on Labour Day to ride around the race oval with the geese in formation.

That winter, while I was out flying on skis over Lake Scugog, one engine on the Lazair quit. I tried a mid-air start. The engine turned over and fired but the propeller did not turn — a broken crankshaft. It was no real problem as the lake was one big frozen landing field, so I made an attempt to nurse it home on the one labouring engine. Gradually I lost altitude. Twin engines on this craft were just double trouble. In the old twin-engined airman's adage, when one engine quit, the good one just carried you to the scene of the crash! This time there was no crash but I made up my mind that the Lazair would become a single-engine airplane.

The next spring we started out again with a new group of goslings. Over the winter I had built a single-engine mount for the Lazair and installed the three-cylinder König engine salvaged from the crushed Easy Riser. It seemed reasonable that one prop would at least halve the hazard for the geese. We never got to test it. On one of the first flights a carburetor float valve stuck, causing the engine to cut out on take-off. Although I got down, there was not enough runway left to stop. I promptly forgot all my basic training and stuck my feet down. Terry had told me this was a dangerous, dumb thing to do. He was right. I was grounded with a broken foot for the critical six weeks that we needed to imprint the birds on the plane, and my hopes for flying in 1987 were out the window.

Six weeks with my feet up gave me time to rethink the whole idea. I realized that a protected pusher prop on the Easy Riser was the only safe solution. I must resurrect the Easy Riser and build in some further refinements to make it perfect for training and eventually flying with the birds.

The trouble was that at this point I had two crunched airplanes and Bill Carrick's enthusiasm was waning. I could not really afford the time or money to take the project another year. I needed a way to finance it. I hit on the idea of making and selling a documentary film of the story — from eggs to airborne ecstacy (I hoped). Carrick liked my plan but was understandably skeptical about the chances of finding a backer.

A few days later I ran into Murray Cooper, an old school chum. He had a large and successful equipment-rental business that mostly ran itself and gave him a comfortable income to boot. He read my proposal and laughed. "Train birds to fly with an airplane? Ridiculous!" But, being the polite fellow he naturally is, he told me that he would look over the proposal with his partner and give me a call.

I met with him a week later. He began by telling me, "My partner thinks I have finally flipped for even thinking about this." My stomach dropped. "But of all the hare-brained, sleazy propositions I've listened to, I like yours the best, so to heck with my partner. I'll go into this one on my own. Let's try it."

So for the third year running Bill Carrick collected eggs from the many nesting geese on his wildlife preserve.

By mid-May the eggs had turned into little cheeping, yellow fluff balls and I carried them down my runway to the new home I had built for them under the wing of an old Easy Riser airframe. During February and April I had totally redesigned the new Riser's landing gear and engine mount and thoroughly taxi-tested it. This was going to be the year. I was bent on succeeding and everything was ready to go.

Out of the blue came an offer to play myself as sculptor in a major 3-D Imax film production, the ultimate in large-format film technology. This would be the second 3-D Imax movie ever made. The shop they wanted me to build as a set for the film and the pieces I was to sculpt were things I had always wanted to do but could never afford. It would mean six months of twelve-hour days to do the film, but it was the opportunity of my career and I couldn't turn it down. I couldn't quit on Murray and the geese either.

Everyone tried to convince me that I should drop the goose project, for it could not be done halfheartedly and it was hard to see how there would be enough hours in a day to do both it and the film. But I could not put the goose thing aside. I just had to do both and somehow get by with less sleep.

I hauled myself out of bed before sunrise every day to feed the geese and spend time getting them accustomed to the sound of the ultralight. The rest of the day and evening was still not enough time to get the four sculptures I had contracted for done and the new shop/movie set built, but I had to believe that somehow it would all work out and I would be ready for the Imax film crew when they showed up in October. My two sons were a great help with the geese and I prevailed on my old Easy Riser flying buddy Jack Weber to help with the new Riser rebuild.

It takes a tremendous amount of patience to work with geese, for you cannot rush them. They take their own time, just like kids, and like kids they need lots of mothering. The more time spent with them, the more they are likely to adopt their surrogate parent completely and accept the strange behaviour of a noisy mother bird with a twenty-eight-foot wingspan.

To get them used to the idea of the ultralight, I would taxi up and down the grass airstrip, with them running and flapping along behind. They were always heading off in some other direction and it took all the willpower I had to keep rounding them up time after time to try again. The giant wings of the Riser constantly spooked the young birds, and I began to think that this really was a crazy plan and that I was wasting my time and Murray's money. Worse, I was jeopardizing my contract with the Imax people.

When the birds began to fly, I really doubted that they would stick with me in the air. The Riser still wasn't completed, and I was pretty sure that in the week it would take to do the work the geese would forget all about me and the Riser and fly off to another life, leaving me and my years of effort behind, crying on the ground. I didn't like it but I had to keep them locked in their pen while we tested the rebuilt Riser. After what seemed like forever, it was ready. It hadn't flown in three years and we had made a number of modifications. I wasn't at all certain the plane would behave properly, and the thought of last year's engine failure and broken foot was still fresh in my mind. High winds over the next few days made even taxi-testing impossible. Each time I came near the penned-up geese they would honk and flap their wings. They were as frustrated as I was.

Finally we got a day with passable weather. I climbed

into the Riser and headed off down the runway, intending to just try it out on the ground first. It popped into the air before I was ready and I nervously chopped the throttle and clunked it back onto the ground, breaking part of the landing gear.

Two days later I was much more careful in the taxi-tests, but in order to keep on the ground I had to use full forward stick. We adjusted the pitch trim. Finally I got it airborne and discovered that the drag rudder stops weren't correct — they returned past neutral and the plane wallowed all over the sky. It was a bit scary to get it turned around and back onto the ground safely. With the stops fixed, it flew great. I took me half an hour or so to get back in the groove of the Riser's strange combined yaw and roll. I did some touch-and-goes. In the redesign job I had managed to save a lot of weight on the landing gear and this paid off in greatly improved climb. The rebuilt Riser cruised easily on a minimal throttle setting. Under power and easing the stick right back, it didn't stall at all. It just mushed along with its nose in the air. With power off, it stalled at just over twenty miles per hour. Full power, stick forward, it did forty-five. Perfect. According to Bill Carrick, the geese should cruise at forty-three miles per hour. I couldn't believe they were that fast after my earlier experiments with the Lazair, but the Riser had the speed if we needed it.

The big day arrived. I backed the aircraft up to the pen. Number-one son Aaron opened the gate. I taxied out with the geese behind me. I got airborne and looked back, but the geese were nowhere in sight. I cruised around and spotted them back on the ground near their pool. Disheartened, I put the Riser back down and got

Aaron to herd the geese out onto the runway again. Again I took off and looked back. They were in the air this time, but turning around to head back to the pool. My heart sank. All this to end up with a swimming pool full of geese! I landed, parked the Riser, and we put the geese back in the pen. I didn't want to talk to anybody for the rest of the day.

I called Murray Cooper to tell him we would try flying the geese next morning and he should come over and shoot some video. At 6:30 A.M. the day was absolutely perfect — not a breath of wind and brilliant sunshine. Murray set up the camera and Aaron stood by at the goose gate while the rest of the family watched from the balcony of the house above.

I preflew the Riser and taxied up to the pen, revving the plane a couple of times to get the attention of the geese. As they burst out of the gate, I hit the throttle and the Riser was airborne in less than a hundred feet. My runway is such that I must make a ninety-degree turn to the left almost as soon as I am airborne, so I could not look back until I was levelled out to the north. When I did, I saw the geese well behind and below, but they really seemed to be trying to catch up. So I throttled back and held it just off stall, flying now over the forest, fifty feet above the trees. Gradually they closed the distance and we were all flying together!

There were geese all around me. I started a little climb and they attempted to form up on my left wing, stringing out in a ragged line. They were getting caught up in the wing-tip vortices (like an aerial whirlpool) and tumbled around a bit. Then the lead goose found the right spot and got almost a free ride. The rest instinctively formed up on him. I executed a long climbing turn and headed back over my house, where Murray had the camera set up.

My head was constantly swivelling back and forth. I couldn't take my eyes off the birds, they were so gorgeous. One dipped down and slid under my left wing and then came up on the leading edge of the lower wing where he found the pressure wave just above the leading edge. For a few seconds he just sat there, wings outspread, surfing freely on the Riser's bow wave. Then he slid gradually across to the right, three feet in front of my face.

It was absolutely thrilling. As we passed over the house at about three hundred feet, my family and Murray were all out there, jumping and waving. A great feeling swept over me as I realized that we had really done it. These beautiful birds were flying at last with their big, awkward mother Riser. It was such a treat to be with the geese in the air.

From then on, I took every opportunity and made any excuse to fly. The geese showed the same enthusiasm. There was something new every time. Sometimes they formed a perfect V off the swept wings of the Riser. Sometimes I was back in the V and watched them strung out in front of me. Once in a while several would surf the leading-edge wave for a free ride. They spoiled the flow over my wing a little, and when the air was still I noticed that the more geese there were on the front, the more power I needed to push us all along.

Like most pilots, I like to use full power to get airborne, and that made me too fast for the geese, who accelerate much more slowly than the Riser. If they got up before me, I ran the chance of actually bumping into them. So I tried to get airborne first and then make a long, slow, sweeping climb. I only flew in the minimal wind conditions of dawn or dusk, and could usually take off to the east where a few open fields afforded me a little more margin in case of engine trouble. The geese climbed slowly behind. When I would approach the pen, the geese would show their eagerness to fly. They stretched their necks up and gave distinctive honks that could be translated into "Let's go, let's go." The honks were interspersed with wing flapping like an engine being revved up prior to flight. When in the air the birds also showed their enjoyment of these first flights. They literally played with the airplane, trying various positions to fly, sometimes surfing on the pressure wave in front of the craft. I maintained about twenty-five miles per hour until they formed up, then we all climbed out over the open fields to the south and levelled off at about five or six hundred feet.

If the air was still, our formation was tight, with the birds tucking in really close. At times one or two would cruise right up underneath me, so close that I could have reached down and touched them. If it got bumpy, the Riser did a lot of rolling and the formation scattered. The geese handled turbulence as if it didn't exist, and easily kept clear of the undulating wings. Most times I carried a little video camera and got as much footage as possible, because every flight was a unique visual treat. I was always sad when we had to return to base because I had used up the Riser's half-hour of fuel.

Landings are the most exciting moments because the geese have a unique way of losing altitude without gaining speed. They pull in one wing and flip over in a tumble and straighten out ten or fifteen feet lower. When they all did it together it was a hoot — like clowning down with a bunch of acrobatic leaves. When they got close to the ground they resumed tight formation and glided in together, looking like a futuristic jet demonstration team.

The most dangerous time for the geese was actually on roll-out after landing. Usually they landed right in front of me, and it was damned hard to steer around all twelve. Only once did I hit them, not one but two at the same time, one wheel over each. I was sick. It looked so horrendous; I thought I had killed them both. Not so. Just ruffled feathers. They were ready for the evening flight.

Occasionally the geese would take a couple of turns around the field before they decided on landing. On one of these approaches, they were gliding in ahead of me when they decided to do another go-round. I said, "You guys can't do that without me!" and I crammed the throttle forward. I caught up with them so fast that I bumped into two. I was freaked, but the geese just did a little roll out of the way, throwing back a glance at me that said, "Where in hell did you learn your air manners, Mom?"

We were a traffic-stopping show when we cruised on a fine summer evening together. There were always several vehicles parked beside the road with the passengers out and gazing up in astonishment. The *Toronto Sunday Star* ran two full-page colour photos on the front and back pages. For the next few weeks it seemed as though every day some TV news crew was filming us. The news travelled as far as California and I got a call from the producers of a show called "Incredible Sunday". ABC flew me and two geese to California to take part. I couldn't help but laugh at the jet-age flight my winged partners were making on an Air Canada 747 so they could make their TV appearance; of course they rode in a dog cage, like any other pet. When we all arrived in Hollywood they walked out on the stage like they owned it. I was worried that they might take off and fly around the studio, but no, they were perfectly mannered geese and did not even poop on camera or onstage.

I marvel more now at the beauty and grace of these birds than I did watching the gulls wheel behind Dad's plow. Watching the geese hour after hour from a few feet away, seeing every feather, every gesture of wing as they cruised through the sky, was an amazing and humbling experience. The experience gave me a new perspective of the air. The geese, or all birds for that matter, are in their natural element while we humans are dependent on some crude technology to keep us up there. We always have nagging doubts about getting down safely, and that little negative feeling limits our ability to enjoy the inno- cent freedom in the air that the birds just take for granted.

Scientists still do not understand exactly why geese and other long-distance migrators fly in certain forma- tions. The classic V associated with geese, some studies show, may not be as important as how close to one anoth- er the birds fly. They can get free lift just from the upward directed airflow around each other's wing-tips by keeping so close that sometimes their wings actually overlap. Some estimates are that in a large flock, energy savings from this can amount to sixty-five per cent, whether or not the birds are in a V, or an oblique or straight line.

The geese stayed in Ontario for the winter while I migrated to Mexico for several months. On my return, I visited Carrick's establishment and entered the pen in which my eleven conditioned geese were wintering. They were among the large flock on the far side of the pond. But when I called, "Come on, geese!" they immediately separated from the larger group and rushed to surround me with much agitation. I truly felt like their long-lost mother.

We returned them to our home airstrip pen in mid-April and took up the flying regime where we'd left off the previous fall. In the initial flights I noted that the geese were more autonomous and would break away to do exploratory flights on their own without the aircraft.

The hierarchy within the group had become noticeably more defined, both on the ground and in the air. On most occasions the same goose would lead the flight, following its own agenda without the aircraft. On the ground it was the same goose which was the most friendly, the one which would rush up to you first, taking a noisy welcoming stance. At the other end of the spectrum were one or two geese that seemed shy or sullen; they always stood off and could not be approached easily. These were always the last geese back in the pen. I also noted that the lead goose had developed a dislike for my daughter Carmen, then five, and would rush at her if she was alone. Poor Carmen was not only frightened but also upset — she had helped raise these geese from cute little balls of fluff; now they were almost her size and one of them was staging attacks on her. Personally, I think that

Bill Lishman

the goose was jealous; it was very friendly with me and showed a great deal of vociferous affection whenever I would visit it. The only thing we could do was to be on guard when Carmen came near the birds. Fortunately, for the most part it was not a major problem.

Weather permitting, I would fly the eleven geese for half an hour each morning, and then they would be left to graze until mid-morning or noon. This daily regimen was carried on through to the end of May. During these "free" hours they would often fly or walk up the hill to our house and hang out on the lawn outside. If the door was left open they would take up residence inside — much to Paula's consternation.

May 28, 1989, they vanished. When I went to put them away at midday, there was no sign of them. There was a great deal of conjecture as to their whereabouts. Carrick's suggestion was that they had flown off to establish themselves in the local marsh in order to go through their annual moult and they would return in a month to six weeks. That did not happen, and after three months I had abandoned hope of ever seeing them again. On October 15, however, the first day of hunting season, three of them returned and landed on the front lawn, the lead goose (a female) and two males. We were overjoyed to see them but made no attempt at that time to return them to their pen. They remained until that evening when, fifteen minutes after sunset, they took off and did not return until fifteen minutes prior to sunrise the next morning. An hour later I took off in the Easy Riser and had my boys scare the geese into the air. Once they were airborne I joined up with them, and they flew with me in tight formation for a fifteen-minute flight, after which they returned to the house. Their leaving after sunset and returning before sunrise became a daily routine that you could set a clock by. I would hear them honking just before their arrival each morning and rush out to welcome them. Twice when they took off in the evening I joined them with the aircraft and followed, trying to discover their nightly rest area and possibly the remainder of the flock. There is a large marsh area to the north of us, and beyond that Lake Scugog stretches fifteen miles to the northeast. They flew directly north past the marsh and out over the lake. Both times I was forced to turn back; flying over the lake beyond gliding distance of land, particularly with waning light, would be foolish.

In the second week of their return I was awaiting their morning arrival, watching the sky to the north, when I saw a large flock approaching, what looked like two hundred birds. As they drew closer, they descended toward the house. In the lead were my three geese, which came in and landed directly on the lawn. The rest of the flock circled twice but were too wary to land, although my three were calling to them exuberantly. Several other times in the next couple of weeks there were similar occurrences; however, only my three would land.

At the end of November we returned those three to Carrick's preserve. In the spring of 1990, on Good Friday, two more of the flock returned, a male and a female, which we also recaptured. Bill Carrick kept those original five geese, some of which mated and raised young. A little later in the spring one other goose returned solo and hung around. I made no attempt to recapture her, thinking she would probably stay for the summer. However, she left after about a week, then returned again in the fall for about a week. She took up the same routine, leaving after sunset and returning before sunrise. The next spring she returned again with a wild mate. They nested in our pond, raised four goslings, and departed. They have returned every year since. She lands in our front yard in mid-March and honks loudly, announcing her return. Her wild mate, a little perturbed at her running home to her parents, usually hangs out at the pond while she greets us for the season. It does not take her long to rebuild last year's nest, and within a week or so she is laying eggs while her mate stands guard.

CHAPTER SEVEN

Making a Living

Despite all our joys and setbacks with the flock, geese were only one element in our lives. Our son Aaron was born in 1972. We bought our wonderful land, one hundred acres, at Purple Hill in 1973 and have lived there ever since. Many of my projects were developed in a one-time barn converted into a shop with all kinds of uses. I was fascinated by some elements of industrial and furniture design, and when something I created in the shop worked well, and would sell, it went into production.

Our second son, Geordie, was born in 1975. Paula, expanding on her textile arts background, in the late 1970s developed an original technique for making knitted fur garments. Based on this invention, she and I as a team developed a line of jackets and coats that became popular in high-end fashion boutiques from New York to Tokyo. In those early days I was the fashion photographer and Paula was the model, two photos appearing in *Vogue* magazine in 1979 and 1980.

By the time Geordie was born, my parents had come

to accept our unorthodox lifestyle. My father just thought the world of Paula. "She was the best thing that happened to Bill," he would tell his friends. In 1965, he sold the farm (except the house) to developers for far more than a handful of beans, and for the first time in his life was relatively well financed. He put money into our growing businesses when the bank rates were at usurious levels. My mother always seemed to support my career as an artist, and took great pleasure in her grandchildren. Paula's parents were also majorly supportive of our endeavours, and I must say, without the assistance of both families during those lean 1970s, our businesses would not have survived.

Since its inception in 1979, the fur business has grown to employ more than a hundred people, with sales in the millions of dollars, both of us active in the management of the company.

At the same time I was working on commissions for such enterprises as Canada's Wonderland, a theme park north of Toronto; Marineland at Niagara Falls; and Expo '86 in Vancouver, for which I did an eighty-six-foot-high sculpture entitled *Transcending the Traffic*.

All of this led eventually to one of my most ambitious and widely praised projects, a full-scale replica of Britain's famous Stonehenge.

It began as an adman's brainchild in 1986. That year, Peter Jarret of Grant Tandy advertising agency needed an original TV commercial to draw attention to the annual automotive ritual of introducing new-model cars, in this case the most exalted high priest Lee Iacocca's Chrysler Dodge Shadow and Plymouth Sundance.

Peter thought that Stonehenge, an arresting image,

might be easily linked to sun dances and shadows. I was appointed the architect to bring that concept to fruition. The task was to create a pagan temple to the automobile god, something like Stonehenge — perhaps out of cars — to use as a setting for the commercial. The imagery of a latter-day Stonehenge made of crushed cars immediately captured me.

I had never been to Stonehenge, but multitudes have and there are a multitude of books. It has been studied forever it seems. My version needed to be the exact scale of Stonehenge, and the location was absolutely critical. The results of our creation were documented later by Christopher Chippendale, a research fellow in archaeology at Cambridge University, who put together an exhibit entitled "Visions of Stonehenge". The opening piece featured a photograph of my *Autohenge*. The same photo was on the cover of his booklet *Stonehenge Observed*, in which he wrote the following:

> The splendidly named Autohenge is a Canadian Stonehenge constructed out of scrap cars, partially crushed in a scrap baler, and tied in pairs with steel bands to make the right balance of height, width and thickness. It was built by local sculptor Bill Lishman in a field at Blackstock in the rolling Ontario grasslands during 1986. It is thirty metres in diameter [about right] and the uprights stand four metres off the ground [about right again]. Its site is superbly chosen; like Stonehenge, it is neither on a mountain top nor in a valley bottom, but placed on a low knoll, complete with a farm track by way of a professional avenue of approach. From a distance its

silhouette so exactly resembles Stonehenge's that a photograph of it will fool most. Close up, the combination of the ancient and grandiose shape with grotesque debased and modern material is startling. It is the most impressive piece of work. Very considerable care was taken in its making to follow the proportions of the ruined original Stonehenge, and the care shows in the feel of it.

In building *Autohenge* there were great moments. I had three weeks to plan it, find a site, locate the cars, erect them, and have the site landscaped and ready to shoot. I took to the air with my ultralight aircraft in the mornings and evenings and flew about the countryside at low level until I had located five potential sites. Then I drove around and inspected them from the ground. The site I finally chose had a magic to it, and when I observed the impact of sunrises and sunsets it worked perfectly. Library research gave me all the necessary dimensions and orientation, so with this information I made a one-fiftieth scale model out of wooden blocks, then set it upon the centre of the site and with a compass oriented it correctly. From that model we laid out the full-scale *Autohenge*. Roy Robertson, who along with his son Sinclair owned the land, got really excited about the project. He also happened to be the local excavating contractor, and took on the job of digging the holes we required for the uprights.

Working in Art Robinson's scrapyard bore no resemblance to quarrying the stone for Stonehenge. It poured rain the whole time we were there, creating a most ugly environment of scrap metal and mud. The cars I needed had to be carefully picked. I could include no Chrysler products and the cars we did choose had to be the right size. Art had a huge old front-end loader that resembled a giant primitive insect. This behemoth would growl and snort down the canyons of stacked hulks in various auto wrecker's yards, searching out the prey we had flagged. Once found, the loader would pick it up in its primitive claws and wind its way back through the masses of twisted steel to the baler, and there drop it. Then like some starving wild boar, its tines like tusks, the machine would tear open the hood and with quick, deft movements break the engine loose from its mounts, roll the car over, stick the forks through the windows, pick it up, and with a couple of momentous jerks, shake the engine out on the ground, snapping hoses and control cables. The machine would then stick a tine through the gas tank and tear it loose. The carcass would be stuffed into the baler. The car would be crushed to about eighteen inches in height, then a second car would be added and crushed to give us a thickness comparable to the stones of Stonehenge.

We handled forty-four cars this way over a three-day period and then trucked them about fifteen miles to the site. Rain had turned the field into mud, so we lost a couple of days waiting for it to dry out. We were ready to go when a crane operators' strike had me stopped again, until I located a privately owned crane with an owner-operator who was willing to risk being blacklisted. (Did the Druids have such problems too?) *Autohenge* was ready on time, and the commercial shoot worked perfectly.

The director wanted to have some birds in the shot, so I was able to bring in the Canada geese that had imprinted on me and followed me everywhere. My original plan was to have them fly in formation with the Lazair ultralight. By the time we were ready to shoot the commercial, I had them trained to the point where they would fly in formation with me and the motorcycle, so when the director wanted a bird fly-by, he got it on cue and in the right place.

A week or so after the shoot, we had a celebration at *Autohenge* attended by all those who worked on it. The geese did a sunset fly-by and returned to land in the centre of the site, not wanting to be left out of the party. We lit a huge bonfire in the middle of the henge. The effect was phenomenal. The light and shadows flickering off all the old cars made them appear like strange, grotesque faces peering into the centre. The final imagery that really capped it was a grand display of aurora borealis, northern lights, that I am sure was ordered by the ancient Druid gods.

CHAPTER EIGHT

The Troubles

If someone asked me to put a title on our flying-with-the-birds experience between 1989 and 1993, the early part of those years would be labelled The Troubles, although they began excitingly, full of promise. After the successes of having the twelve geese fly with me in 1988, Bill Carrick talked much about adapting what we had learned to the cause of the trumpeter swan. When I was a child, my mother had told me of the plight of many species of endangered birds. She told me the story of the demise of the passenger pigeon; that there were fewer than thirty whooping cranes left in our world; and that the trumpeter swan was severely endangered. A great deal of effort by members of the Trumpeter Swan Society had gone into attempting to take breeding stock from the Alaska population and re-establish flocks in their traditional habitats across the continent. However, these transplant-ed birds had little or no luck in establishing new and safe migration routes, that knowledge having died out with the demise of the original flocks. In 1989 Carrick, a member of the Trumpeter Swan Society, had planned a presentation at the society's biannual convention in Minnesota. That fall, he and I flew to Minnesota and showed film of our goose endeavours of the previous year. The presentation was met with everything from ridicule to full enthusiasm. The most enthusiastic supporter was William J. L. Sladen, professor emeritus from Johns Hopkins and the head research scientist at the Airlie Center near Warrenton, Virginia. He is an impressive figure with a forceful manner of speaking in particularly precise English (he was born in England) and with a nose that interested the sculptor in me — a human adaptation of a swan's bill. Later I found that he was more than just your run-of-the-pond ornithologist.

As a schoolboy in England, he would always win the prizes for best butterfly and pressed-plant collections. The natural world remained appealing to him as he trained for a medical career, working in a plastic surgery unit with the burn victims of Montgomery's tank war with Rommel's Afrika Korps. He became fast friends with the late Sir Peter Scott, a noted waterfowl artist, ornithologist, and son of the famed Antarctic pioneer. In 1947, with the British equivalent of an M.D. and Ph.D. in medicine, Dr. Sladen signed on as medical officer for the first of several Antarctic expeditions he would make. His research there on Adelie penguins later earned him a Ph.D. in zoology from Oxford. In the 1960's, he discovered significant levels of the pesticide, DDT, in seals and penguins in the Antarctic; a historic revelation, as it alerted the world that modern chemical contamination had spread throughout global ecosystems.

Dr. Sladen's interests spread to swans and geese. He

pioneered banding and radiotelemetry techniques to track their migrations, and was the first western scientist invited by the Soviet Union to conduct research there on migratory waterfowl. During a frantic few months in the spring of 1973, he and his team used small planes and rented automobiles to follow migrating tundra swans on their northward route from Chesapeake Bay toward Arctic nesting grounds — the first time anyone had done it. The swans on that trip easily outflew the plane, which would land to refuel while the tracking team followed the swans along the Ohio Turnpike and other roads as best it could.

His many recognitions include a Member of the British Empire from King George VI and the Polar Medal from Queen Elizabeth. There is even a mountain in the South Orkney islands named after Dr. Sladen. He is a respected power in scientific organizations seeking to reintroduce trumpeter swans to their old migration habits that might help them recover from the near-extinction.

All of this might sound a little stiff and formal. In fact, Dr. Sladen, though every inch the scientist, retains all of a schoolboy's passion for the natural world, especially an utter fascination with what he refers to as the "macrobirds". To him, the large waterfowl like the swans are magnificent ambassadors. Their spectacular comings and goings can stimulate a greater public appreciation of the need to preserve wetlands and larger landscapes that are critical habitat for the birds. Re-establishing the old migrations, and even new ones, for trumpeters and whooping cranes, would help the species, and also help in preserving a series of refuges and resting places along the migratory routes.

Dr. Sladen, captivated by what we had achieved in persuading young geese to fly with my aircraft, invited us to fly a flock of swan from Ontario to his research facility in Virginia. This fit in exactly with what Bill Carrick wanted to do, and I was hooked as well. "What a wonderful application of what I've learned with the geese." I thought.

Of course, before we could set off for Virginia with a flock of swans in tow, a lot of work had to be done. Our first run at this project began that same fall, 1989, on Lake Scugog. Bill Carrick had three swans conditioned to fly with a boat — left-over stars from a 3-D Imax film of the previous year. We would attempt to get his swans, already imprinted on a boat to fly with an ultralight as the geese had done. I acquired a new Cosmos aircraft and fitted it with floats. However, the three swans refused to have anything to do with forming up on the aircraft, apparently spooked by its big wing. Then there was an early freeze-up that covered Lake Scugog with a layer of clear ice thick enough to walk on. Carrick and I had one hilarious (in retrospect) learning experience, slipping and sliding on foot over this minimal ice layer trying to recapture those three swans, which could not take off and fly out of there because doing a take-off run on ice was not something they had perfected! So we slithered after the swan, probably the closest I will ever come to the sensation of walking on water, but in the end it was all in vain.

I thought later that if we'd tried harder, if we'd been able to work at it longer, the result might have been different. But there are only so many days when the flying weather is right for this kind of project. So we had no success with the swans that fall, but in our post-mortems we decided to try again the following spring with a new plan we thought might work. This would be to get just-hatched swans and imprint them on me and the Cosmos aircraft, the system which had worked with the geese.

Meanwhile, we made a weekend trip to Airlie to have a first hand look at Dr. Sladen's impressive operation: three thousand pastoral acres in the rolling hills of Virginia about fifty miles to the west and south of Washington D.C., interspersed with ponds, woodlots, and old done-over Virginia manor houses. Created by Dr. Murdoch Head as a medical and environmental research centre, it is also a wonderfully picturesque and popular rural convention facility.

I knew that this project needed people — including other pilots who could fly the Cosmos. In February 1990, five of us gathered in Mexico to take advantage of good flying weather and become familiar with flying trike-type aircraft like the Cosmos. Quite different from a standard stick-and-rudder craft, they are controlled by weight shift and flown like a hang glider, a very simple airplane with few things to go wrong and easily foldable.

The group in Mexico included myself, Joe Duff, Bill Carrick, Murray Cooper, and a freelance journalist in Ajijic Mexico. With barely fifteen hours of trike time, I was the designated instructor, and although everyone got several hours on the trike that we rented, the only one to go solo was Joe, a long-time pilot with a natural feel for the air. We did have a lot of fun. Bill Carrick, at seventy and in Mexico for the first time, was fascinated by the different flora and fauna and kept us expertly informed of the Latin names of what we saw. Once we were parked outside a small store, and while Joe ran in for some refreshments, I picked up a perfectly round dried horse dropping from the street. I passed it to Carrick and asked him to identify it for us. After careful study he declared it to be some kind of fruit. I informed him he was correct and that the common name in Canada was "road apple". Furthermore, when frozen, it was used as a puck in road hockey, but I wasn't sure of the Latin name. Maybe *equus pommus?*

At this point, the spring of 1990, the trumpeter swans were Bill Carrick's project. With his experience in such matters, he was to deal with the bureaucracy in the United States and Canada, getting permits and so on, and I was to get the aircraft operational. We worked hard. I took on the task of imprinting a flock of Canada geese, while Murray Cooper, who had several ponds at his home, would take on imprinting the new swans when they arrived. As time went on, Carrick had a great deal of trouble with the wildlife officials whose permit we needed. They seemed to think our plan was too bizarre to treat seriously, creating a level of frustration that included frayed tempers. I was pressing Carrick hard for more action. At one point he told me, "You're taking over my project," and I replied, "No, Bill, I'm just trying to get things done." As we debated several points of conflict, I said the fatal words: "You can't do this without me. You need me." At that point, Bill Carrick and I started going our separate ways.

When I discussed the situation with Dr. Sladen by phone, he told me that there were seven swan eggs in a nest at the Airlie Center, almost ready to hatch and likely to become prey to snapping turtles. If we wanted them for the all-important imprinting when they hatched, we should pick them up immediately. I flew to Airlie, got the eggs, and then found that to legally move the eggs to Canada, we had to drive to Richmond, Virginia, to get the official O.K. from a state veterinarian. This we did. Back in Ontario, five swans hatched on schedule. In the long weeks to follow we worked at the essential imprinting, both of the swans and of a flock of goslings. The plan was to try some cross-country experiments with the geese first. The swans, which begin to fly much later in the season, usually September, would be ready by the time we had successfully flown the geese for a few weeks. I also thought that sometime — not this year if the swan project worked, but sometime — I should fly a flock of geese to Airlie, leave them there for the winter, then see if they would successfully carry out the essential second part of the migration pattern by flying back to my place the following spring. This was just an idea, its time not yet come.

With all that on my plate and also working on building a new radically designed family home, our Earth House (built into the top of our hill), I didn't think anything else could possibly happen to interfere with our long-planned swan flight to Airlie.

I was wrong. First, I felt that the training of the swans was not going as well as I wished. Also, my new flock of eighteen Canada geese required constant work if we were to attempt their cross-country migratory flight experiment, a prerequisite to trying it later with the trumpeter swans. Raised in a similar way to my earlier flock, these geese were well on the way to being trained when in early

July we were shocked to discover that the Canadian aviculture permits that both Carrick and I were working under were not considered valid by the Canadian Wildlife Service. Our permits were suspended and the experiment halted by the service, more specifically by one bureaucrat, Joe Carreiro, who seemed to have a long-standing disrespect for both Bill Carrick and Dr. Sladen. Carreiro was in charge of enforcement and compliance for the CWS. He also issued permits for scientific purposes. On June 22, Ministry of Natural Resources officers first visited both me and Murray Cooper. Simultaneously other MNR officers arrived at the premises of Bill Carrick and his associates, Janet Huen and John Brouers. In that June 22 visit to me, Officer Al Giesche, a special services investigator, wanted to see our aviculture records. These were produced for him. I also told Giesche that the swans were a hybrid cross between trumpeter and tundra swans and had been legally obtained as eggs from Dr. Sladen. I wasn't worried about legality at that point. Before picking up the eggs from Sladen, I had asked what documents were necessary to import the eggs to Canada. I'd been told by an official at Agriculture Canada that a U.S. federal health certificate would be required, nothing more. I'd secured that, so I was under the impression that the eggs had been legally imported into Canada. I also informed Giesche why the swans were required — for the important migration experiment in which both Canadians and Americans were involved.

I might as well have saved my breath.

Giesche, along with another officer named McQuay, pronounced the sentence: the swans were being seized, and were now the property of the Crown. They gave no reason for the seizure, but for the sake of the birds' health they did not remove them physically at the time. They acknowledged that the birds were being well cared for but ordered that they were not to be removed from Cooper's premises. Under threat that the birds would be physically taken if we did not comply, I signed Officer Giesche's notebook acknowledging that the birds were under seizure.

On or about June 28, after I had consulted with Carrick and Harry Lumsden, a retired biologist from the Ministry of Natural Resources and noted trumpeter swan specialist, I called Officer Giesche and requested that the seizure be lifted. Giesche's reply was that none of our group knew anything about law enforcement and the swans were still under investigation.

About 8:00 A.M. on July 5, Officer Giesche and another unidentified man arrived at the residence of Murray Cooper and without showing identification or papers of any sort, physically took the five young swans and put them in the back of a rental van.

When I asked who was seizing the swans, Giesche's reply was, "The government."

I asked where they were to be taken. The answer was simply, "to a facility."

I asked if there were any charges. The answer was yes, perhaps conspiracy. This seizure procedure we documented on video. The two men left with the van and swans, refusing to reveal their destination. My son Aaron and I followed them. The chase halfway across Ontario had some humorous aspects. They kept trying to lose us on back roads, but didn't. All that time we were getting very low on gas and were afraid that if we stopped to fill our

Bill Lishman

near-empty tank, they really would lose us. But we were still with them when they turned into the Ministry of Natural Resources office at Hespeler.

When I got out of the car, our gas tank showed empty.

I asked Giesche again why the swans were being seized. He told me that if I did not stop following them, I would be charged with obstruction and that he would call me later in the day.

No phone call was received, so I phoned Giesche at 7:45 A.M. on July 6 to be informed that there were several charges pending:

1. That the swans in question, hybrid or not, were classified as migratory birds.
2. That the birds were exported from the United States illegally and that no U.S. export license had been obtained.
3. That Dr. Sladen had no licence to deal in migratory birds and did not have a United States export permit for migratory birds.
4. That while I had an aviculture permit, the birds had been at Murray Cooper's place, so Cooper really had possession and did not have a permit.
5. That we — Carrick, Cooper, Lishman and Sladen — intended to release these birds to the wild and that was illegal and therefore we were conspiring to do an illegal act and might be charged with conspiracy.

Murray Cooper was charged with possessing migratory birds without a permit. I felt this was all very autocratic and pursued the matter on a live radio show (CBC's

"Radio Noon"), in which the host, Christopher Thomas, helped bring about an arrangement for me to meet with Carreiro at Ottawa a few days later. There, Carreiro just waved his finger like Big Brother and told me how idiotic this whole idea of the flight to Airlie was. He claimed that an ultralight aircraft could not fly that far and that I would never do it. They kept the swans, which I argued against and lost. They also ordered me to stop flying with my geese, which I ignored.

But by that time even I had to admit that the swan project was legally dead. We could launch a court action against the Crown only at prohibitive cost. It was time to stop. I did wheedle an extra month of flying with the geese, but one more disaster struck. Early one morning a neighbour's dog broke into the goose pen and killed nine of the eighteen geese. It took over a week before I and the traumatized nine surviving geese were able to fly together again.

It was a sad autumn. Feelings were strained on all fronts. It seemed that our wonderful plan to help re-establish migratory patterns for swans was impossible. As for the five swans, they were returned to Dr. Sladen after the CWS had arbitrarily operated surgically on the young signets to render them flightless. The charges against Murray Cooper were dropped.

CHAPTER NINE

Temporary Burnout

The combination of these setbacks in 1990 and the strain of designing and constructing our new Earth House left me exhausted. By the fall of 1991 I had no energy for anything, no enthusiasm, and a constantly upset stomach. The thing I did best was sleep. I thought it was some kind of virus. I went to my doctor for a check-up, but he could find nothing wrong. He sent me for test after test — blood tests, barium through the bowels, radioisotopes through the liver — but nothing definitive turned up. I gave up and went to a well-known homeopath. I sat down with him for a couple of hours. He asked me no questions, took a little blood sample, checked my pulse, spent half an hour inspecting my irises, and then gave me a list of my symptoms. Basically, he said, I was severely run down and the best route back to health was diet, this group of minerals, these vitamins, and physical exercise every day. I went home and did as he said. In a couple of months I was feeling almost human again, though it was six months before I had enough energy to attempt any further projects.

In the fall of 1992 I got a phone call from a Wisconsin businessman who was a supporter of the whooping crane restoration program and therefore on my side, urging me not to give up. This was followed by a letter from George Archibald of the International Crane Foundation. He asked if I would be interested in doing some migration experiments using ultralights with cranes. We talked about that in several phone calls. He explained how tenuous the existence of the one surviving flock of whooping cranes was. According to a book I read, *Reflections: The Story of Cranes*, biologists estimated that in 1865 there were 1,200 to 1,400 birds alive, "but by 1890, the whooper had disappeared from the heart of its breeding range in the north-central United States." In some respects it was an old story of steady decline, breeding grounds altered and disturbed as settlers broke the prairies and drained marshes for farming. Whoopers also were hunted and their eggs collected as specimens. By 1941, a small Louisiana flock had been all but wiped out by a hurricane. Fifteen or sixteen whooper wintering on the Texas coast of Aransas National Wildlife Refuge were all that remained.

Where the whoopers summered, however, was a mystery. The species seemed destined for extinction, joining the passenger pigeon as birds commemorated on statues and fated, as Aldo Leopold put it, "to live forever by not living at all." Then in 1955, at Wood Buffalo National Park in the Northwest Territories, personnel returning from fighting a forest fire noticed a whooping crane family. The breeding grounds had finally been discovered! This last flock, which bred in Canada and wintered at the Aransas refuge in Texas, became the focus of long-standing cooperation between the United States and Canada.

The migration route it follows, stretching some 2,500 miles over land between Canada and Texas, is a perilous one, however. Along the way, cranes are exposed to inadvertent shooting by hunters. The immature young are not easily identified as whoopers like the parents, and the cranes' strong family ties means that the adults and siblings tend to stick close to a downed bird, also making them vulnerable. Collisions with power lines have also taken their toll over the years the crane migration has been observed, and in the western states across which cranes pass there is the danger of settling for a rest in mining areas where the water is deadly acid. Then there is the vulnerability, both physical and genetic, that comes when your whole race is concentrated in a single group and place. An extraordinary hurricane, a siege of disease (apparently responsible for killing eleven of the whoopers in 1990) — even an explosion on one of the petrochemical barges frequently passing the Aransas refuge — any of these could end the existence of earth's last whooping crane population. Some scientists fear numbers of the cranes are already so reduced that they will, in the long run, lack the genetic diversity to adapt to changing environmental conditions. Present plans at the Patuxent Wildlife Research Center are to try to build the flock up to a minimum of 500, which managers there think will afford at least a minimally adequate genetic base for long-term survival.

Dr. Archibald pointed out that if a second flock could be established and taught to migrate a lesser distance, it would be at least a backup to the 130 or 140 individuals that made up the existing flock.

The magnitude of such a project, while it would build on our work with the geese, would have to go well beyond it. For example, migrating geese may well find good stopping places along the route by keying on other flocks of geese that have set down for a rest. If you are a whooping crane, there simply are no other whooping cranes along the way. Places along alternate routes would have to be sought out, and the cranes taught to use each one.

Still, it seemed worth the effort to give back to people across North America the sights and sounds of such a bird. The whoopers are one of the world's gaudiest and most elegant dancers when in mating mode. With sweeps of their great wings, they bow and curtsey, leapfrog one another and tumble head over long shanks; they leap, arrowlike, into the air. Even on their winter feeding grounds they are a presence, as an old description from an 1883 magazine *Forest and Stream* testifies:

> In fine, calm weather, [the crane] delights to mount up, in great, undulating spirals, to the height of a mile or so, and take a quiet float, while he whoops at neighbors in the adjoining counties. After airing himself to his heart's content, he descends, sometimes spirally as he rose, at other times with great plunges and wild, reckless dives, until within about 50 feet of the earth when he hangs himself upon the air, with his long, spindling legs down, gently settles and alights.

I spoke about this project with Dr. Sladen who suggested we all meet at Airlie in Virginia in December

1992, which we did. We all agreed that we needed to carry on the ultralight/migration experiment in some form, but how, and with what focus?

One push was for a first experiment that would involve sandhill cranes, perhaps a flight from northern Michigan to Louisiana. Dr. Sladen's main interest was in trumpeter swans being re-established in the east. Also at the meeting was David Ellis from the U.S. Fish and Wildlife Department Patuxent Research Center, who had been experimenting with imprinted sandhill cranes that would follow a truck. There were questions of how any experiment like this would be funded. Dr. Sladen talked of getting funds from *National Geographic*. George Archibald felt we could raise money through the International Crane Foundation. Dr. Ellis, though employed by the U.S. Fish and Wildlife Service, could see no help coming from that area. In fact, getting permits to carry on an experiment as risky as we envisioned would mean having to go through another tangle of bureaucracy and run a good chance of being vetoed.

By the time I left Airlie I was totally confused as to any direction that could be taken. I went home and slept on it and within a week came to the conclusion that the only route to take was to try another experiment with Canada geese. If the experiment went awry it would not jeopardize any endangered species. I knew enough about geese and flying with them to improve the likelihood of success. I could not see any way of having a third party fund the project, since it involved both Canadians and Americans. I could find no international funding vehicle whose criteria covered this experiment. My conclusion was to treat it primarily as a scientific experiment but to fund it as a film production. One would not jeopardize the other. In my mind it would be a good symbiosis though a gamble. If the experiment failed, so did the film. My half-hour video *C'mon Geese* about my early experiences with geese had been modestly successful financially. I kept getting letters from people who had fallen in love with the story and were ready to purchase the next instalment. It seemed the only way to go. I made phone calls to all concerned and they all liked the idea, but there was a worry about making a commercial film from it. There is a school of thought that believes science remains pure only when there are no financial rewards that might tend to warp the outcome. Secondly, there was a perception among some scientific groups and bureaucracies that the film industry exploits animals. Finally, if we made the project into a film it might be misconstrued as just a film, not a proper experiment. Those were the negatives. My own thinking was that science exists for the enlightenment of man, every man, and that we were using film and video as the appropriate, up-to-date way of recording our findings.

The experiment, simply put, was to raise a flock of Canada geese as I had in earlier years, imprint them on an ultralight aircraft, and fly them south in easy stages, four hundred miles to the Airlie Center. Then we'd see if they would return to Ontario the following spring on their own. If not, we'd fly them back to our home base at Purple Hill and see if they would return to Virginia the subsequent fall.

Covering all the bases single-handedly would be impossible. While Dr. Sladen had agreed to write up the scientific proposal and would do a lot of the political background work required, I needed a hands-on dedicated person plus a complete backup in case I, through some mishap, was removed from being active. In previous years, I had relied on my children to help, but they were older now and growing bored with my goose projects.

The person the project needed would have to be comfortable with many different demands, and also have that indefinable talent to relate to the birds, not to mention being an excellent pilot of ultralight aircraft. The right guy was Joe Duff, a long-time ultralight pilot and friend. He had watched with great enthusiasm from the sidelines during the goose flights of 1988 and 1990. It took little persuasion for him to set aside his successful commercial photography career (another talent that proved essential to the project) and take on this adventure.

We would fund what we called Operation Migration through profits from Paula's business of designing and creating fur fabrics and clothing. It was generating good profits and if the project failed, the loss would be a write-off.

Just as I made the commitment to carry on with this next experiment on that financing basis, I was thrown an unexpected curve. The fashion business carries Paula's name and is based on her invention of knitting fur. As the company was solely owned by Paula and myself, we had many a late-night directors' meeting. Normally, I had little to do with the day-to-day running of the company. The problem was that Paula Lishman International had grown so large, with more than one hundred employees, that it had outgrown the management. The current manager, one of the company's first employees, had with great dedication worked her way to the manager's chair, but the job had seemingly grown beyond her capabilities.

The rumblings and grumblings throughout the organization had grown to a crescendo. Many important matters were falling through the cracks. It was affecting the morale of the company severely, jeopardizing its ongoing well-being. Paula, who was president, chief designer, and marketer, was completely overstressed, and getting ready for a three-week sales trip to Japan. With too much on her plate already, she could not take on the day-to-day administration of the company. The only answer was for me to jump into the manager's chair until we could get someone to replace me and take over the reins.

At first, I attempted to get the current manager to realize the situation and initiate some new programs, plus delegating a lot of the workload. After a month we had no choice but to let her go, which was one of the toughest things that Paula and I ever had to do.

There was another factor: I had started a large sculpture in late November. It was not a commission, just a piece I wanted to do. I had spent many hours in November and December doing the clay studies and my long-time assistant, Richard VanHeuvelen, and I had started on the full-scale piece before the year's end. It was to be a twenty-two-foot-high pregnant female.

Now I have to tell you why I was inspired to do it. The previous year, an interesting woman, author Camille Paglia, had been interviewed on a local television station and had spoken about feminism in a way I could understand. Like *vive la différence!* I was so fed up with so many people getting equality mixed up with sameness. Paglia stated that males and females are two different species inhabiting the same planet. I do not go as far as to say that, but I certainly do believe in the differences and that one sex can never quite understand the other. It is this that keeps us interesting: to me, the more difference, the more intriguing it is. Anyway, my wire-frame pregnant sculpture was to be a statement. I would make her twenty-two feet in height, which would make the foetus in the sculpture the same size as an adult human. The mother would be in wire frame, transparent and the foetus would be solid.

So now here I was managing a company of one hundred women while also attempting to create a twenty-two-foot-high sculpture of a pregnant one and planning to fly south with a bunch of geese in the fall. Something had to give. The sculpture was put on hold. I would carry out the migration experiment, and replace myself as manager of Paula Lishman International as soon as possible while trying to get things reorganized in the company. There were times when I felt I had taken on the twelve tasks of Hercules, of which the most difficult and urgent (to me) was getting back to the migration project.

CHAPTER TEN

Trying Again

Operation Migration could not really get started until we were sure of receiving the necessary permits from the Canadian Wildlife Service, which seemed inordinately long in coming. Dr. Sladen's enthusiasm never waned. He was putting together the scientific proposal, the key to getting permits in place, although by then we were not even sure what permits were required. I had written in 1992 to the CWS as I had the previous year, hoping in vain for action on permits.

My 1992 letter had to get their attention so we could get their approval in time to start. I began my letter to Steven Curtis of the service by explaining that Stephen Low Productions of Montreal had been commissioned by Sony Japan to create an educational yet entertaining wildlife film using their newly developed high-definition television system. It would be showcased internationally, a very prestigious commission to go along with the awards Low had already won. I explained that he wanted to work with me on the concept to graphically capture the massive migration of the sandhill crane and/or the snow geese. In politely requesting a permit to raise a flock of both species here in Ontario and to condition the birds to fly with the ultralights, I also asked for his recommendation of a biologist or a group of biologists who would be knowledgeable about migratory patterns and would be interested in working on this project. The film project could be used as a vehicle for furthering post-graduate research. I stressed that the deadline for hatching the birds was tight and crossed my fingers.

My answer to this letter had come in May, too late to even consider a start that year. Now, February 1993, I wrote to Steven Curtis again, supplying him with two copies of a scientific proposal to conduct a migratory bird experiment and support letters from several eminent ornithologists who had been consulted in the preparation of the proposal. My letter pointed out the far-reaching affects should the results of this experiment be positive and that what we discovered might be key to the restoration of the whooping crane. We were already receiving intense focus from the members of the International Crane Foundation and Japan. I referred to the bad start I had had with the CWS and how I was sure they would see that now I had the right backing and guidance from qualified scientists. The hard part of this letter was to address the service's lack of timely response. If you're smart, you don't tell a bureaucracy they messed up, but I wanted to point out that we had missed an opportunity the previous year because my letter "fell through the cracks". I stressed that I would be happy to meet with Curtis or anyone else at CWS who had concerns, at their convenience, and would call him if I hadn't heard from him by early March. And I ended the letter by mentioning that Dr. Sladen had already presented the project to the Atlantic Flyway Council and was in the process of securing the necessary permits from the various states we planned to overfly.

What we needed most was to soften the CWS's negativity toward the project. Under pressure, they gave a bit at a time. The main problem seemed to be persuading Joe Carreiro. A friend of mine overheard me speaking on the phone to him at great length and complimented me on my patience, patience not being one of my strong points. Carreiro then issued the permit in stages. First, a permit to collect eggs came in mid-March 1993. Then an extension to do preliminary imprinting. The initial permits specifically forbade flying with the birds if other permits from the United States were not in place by certain dates. This kept us off balance, making our full commitment to the project a gamble still, but finally we had the permits we needed, and in early April Joe Duff and I began our egg collection. Our first target area was the harbour at Whitby on Lake Ontario which was literally infested with Canada geese. At first we were a little apprehensive about raiding nests, for it seemed hardly fair to the geese. And because the harbour was a fairly public place, we were a little concerned about how the locals would react to our egg thievery, even though we carried permits which made it all quite legal. We needn't have worried.

The geese were nesting around the edges of the harbour, near an old boatyard with a number of rusting barges, piles of dirt, and so forth, hardly a natural setting. When we began we could see several pairs of geese from

Bill Lishman

the roadway, but the boatyard was fenced and locked, and there was no one with a key. It was a Saturday. We had received permission to go in but the only way was to scale the fence. We felt as though we were breaking and entering, then trespassing, as we picked our way through last year's dried weed, old cables, and pieces of scrap marine hardware.

The first nest we spotted was on a mound of earth at the water's edge. If I had been a kid I would have built my medieval castle on top just where the pair of geese had placed their nest. As we approached the nest the female sat on the eggs, her neck and head held low and absolutely still, hoping that we would not notice her. The gander was in the water about twenty feet offshore, alert to our every move, head held high, swimming in tight circles.

I got up to the foot of the mound and waved my hat at the sitting goose. She of course realized her cover was blown and reared up, hissing, her wings held wide. The male took to the air and fluttered around her, landing back in the water, a somewhat awesome display of power. The goose's position on the mound put her several feet above my head. I really felt disadvantaged. Then some old Viking instinct took over. I boldly strode at her, waving my hat, Joe backing me up. Again the gander came in from the air hissing and hovering in mid-air. I gave out my best attack yell and continued the advance up the mound, striking out with my hat. The geese backed off, both landing in the water, honking valiantly at me. I gingerly picked up the eggs, one at a time, while keeping vigil that the geese did not stage a counter-attack as several lovely warm eggs were put carefully in the lined satchel brought along for the job. As I retreated down the mound with the booty the male staged another attack, but I was able to fend him off again with several swipes of my hat.

We found about five other nests in that old boatyard and really got a good start. Finding them was the real trick, for a mother goose can hide a nest so well that you might easily miss it while staring right at it. Joe and I walked within two feet of one that was right out in the open. I passed it by and so did Joe, but he turned back, saw it, and tapped me on the shoulder. There was the mother goose as still as her surroundings, barely a foot or two from us.

For days we slogged through marshes and swamps searching out those elusive nests, following chevrons of geese in the air to find out where they landed. My fears of being accosted by locals when they saw us raiding nests were groundless. In fact it was quite the opposite, for Canada geese are so abundant that most local residents have had quite enough of them greasing their manicured lawns. We visited one private pond where we were told a pair was nesting. After an arduous search in which we tromped the total perimeter of the pond, the only thing visible was a half-submerged log at one end. I just happened to be looking at the log when I noticed the slightest movement. Getting closer, the "log" turned out to be two adult geese stretched out together, mimicking a floating log. When we got right up to them, there were four goslings hiding behind them. We were amazed at how they had been able to fool us.

Back home with our eggs, we set up the commercial incubator we had acquired. Joe produced a wall chart to record temperatures, humidity, and egg turning, which became a ritual. We candled the eggs at the appropriate times and eliminated a number of infertile ones.

The week before the egg hatching, Joe and I flew to Dijon, France, to meet with Gérard Thevenot of La Mouette and test-fly one of his new wing designs that he thought might meet our goose-speed requirements. We spent a few days at Cosmos/La Mouette testing the speed range of the new sixteen-square-yard double-surface wing called The Ghost. At that point there was not a trike in existence that we knew of that could be mated with that wing to give us the right speed envelope and range. Gérard concluded that he could modify the wing to allow the slower speed that we required, when it was mated to the Cosmos single-place Echo trike.

We arrived back home two days before the first eggs hatched. The goslings, wet and bedraggled at first, could hardly lift their heads, but within a day they were lively yellow fluff balls ready to take on the world

From then on it was a dawn-to-dusk labour, playing parent to the quickly developing goslings, working out the next stage of permits, organizing photo and video technology, building pens, making a mock-up aircraft and a second hangar. Our crew had grown to four. My son Aaron, at first hesitant about working with geese for yet another season, was pressed into service, and Richard VanHeuvelen, who had worked with me for the past fifteen years, found himself more and more involved in the project.

Joe and I were in constant touch with La Mouette and Cosmos on the development of the wings. Gérard flew them. His report was a little disheartening. It did not sound as if the trike we had planned would fly slowly enough in its standard configuration. My experience with König radial engines had been great. The Easy Riser had flown several hundred faultless hours. While putting out only twenty-eight horsepower compared to the fifty from the ninety-pound Rotax engine that the Cosmos came with, the König was half the weight, had an electric starter, propeller, and all. We only wanted to fly at thirty miles an hour anyway, and the König sipped gas at only seven litres per hour. The geese now were turning from fluff to feathers, starting to look like real geese. I ordered the trike from Cosmos and the engine from König and the wing was en route from La Mouette. Nothing showed up on time and we were forced to imprint the geese using a hastily built plywood mock-up aircraft on the ground.

By July 1, the geese were fully feathered. They had developed their pecking order and their different personalities began to emerge. It is so important to play the parental role and spend time with the growing birds that usually either Joe or I would sit in the mock-up aircraft on the runway, surrounded by the young birds, laptop computer clicking away with necessary correspondence and the portable phone in hand.

In early July we drove the route to Airlie in Virginia to locate potential landing spots along the way. At first we considered state parks or conservation areas, but thought again about dealing with another level of bureaucracy and opted for checking the possibility of using private grass airstrips. I contacted the Experimental Aircraft Association in Wisconsin, got a list of the local chapters, and through them identified a number of potential airfields. It took a week, but by the time we returned from Virginia we had found eight private airstrip owners en route more than willing to cooperate.

The Echo trike we had ordered from Cosmos came as a bare-bones box of parts with no assembly instructions. Within days, we designed, built, and static-tested the engine mount and throttle-control system and also put the engine through its break-in procedure. Every hour or two we would stop and spend some time with the geese, trying to get them to follow us up and down the runway while pulling the dumb-looking mock-up. We experimented with splitting them into several small flocks in an attempt to maintain our positions as flock leaders. This meant a fever of building more pens.

By July 13, the first aircraft had been run on the ground for two hours, everything checked and rechecked. I test-flew it with the Ghost wing from Willy Casteel's 3,000-foot strip on Scugog Island, as my rough 900-foot strip left much to be desired as a location for test-flying. The Echo flew perfectly from the start. The climb rate was adequate but certainly not like the climb of the sixty-five horsepower two-place, which rockets out at such an acute angle that it is scary to use full throttle. The more standard rate of climb of the König-powered Echo commanded a little more respect. Pitch control was fine, but like most double-surface high-performance wings, it took some muscle to control the roll. The air-speed check seemed O.K., but hard to read, for the wind meter had to be repositioned according to the cruise attitude of the wing. We found it could not be flown at goose speed without a great deal of strain.

Next we tried the plane with the geese. The first flights were extremely disappointing. The geese, in several groups, flew differently. Some would not fly more than a few feet above the trees and some would even fly down among the trees. Others would land immediately. None, at first, would come anywhere near the aircraft. The first few weeks were terribly frustrating and wearing on Joe and myself. For the first time in my life I got up in the mornings secretly wishing the weather was too bad for us to fly. We took turns trying to get those birds to form on the aircraft in the air. With their excellent camouflage against a summer forest, it was easy to lose sight of the geese from above. The forest huggers were the worst, for our safety factor hinged on engine reliability. Fortunately there was never any indication of engine problems.

The König has an independent magneto for each cylinder and will still climb on three of the four cylinders, maintain on two, and keep running on one; plus, we had installed a secondary electric fuel pump running parallel to the engine pulse-driven pump. We could switch on this pump if we were flying in low altitude situations to round up our wayward flight partners. We kept telling ourselves that this set-up was safe, but it still was a constant strain. In our heads was the drill to follow in case of a tree landing. Simply put, the rule was in conifers, wedge it in, and in deciduous, stall it in. For three nerve-wracking weeks, morning and night, unless the weather prevented it, we worked at getting those birds to follow in the air. They were fine on the ground but it took them the longest time to realize that we wanted them to fly with us in the air. Besides that, they landed time and time again in the strangest spots. We spent day after day searching them out and walking them back to the pen. We were on the verge of calling it quits. Instead, we took another tack and made some modifications to the aircraft. It began to work. First two, then three geese would fly with us. We kept adding birds all the way up to eighteen.

It looked now as if the flight south to Airlie might work after all, but a new set of worries kept us up nights. Our first leg meant crossing almost forty miles of Lake Ontario. Would the geese follow us over that great expanse of water? What would we do if they got tuckered out and decided to land halfway across? How could we minimize the risks if we had an engine failure and had to ditch in the cold water? How would we arrange to clear Immigration and Customs on the U.S. side? How would we get accurate, up-to-date weather reports? The last question was answered by Met Tech, a private weather-consulting firm in Montreal. Its people agreed to furnish us with detailed briefings during the weeks prior to take-off and as we journeyed south.

The flying now became enjoyable and by early September we had the second Echo trike airworthy. Joe and I found that flying one fast wing and one slow wing was the right combination of goose leader and goose chaser. The second Echo was powered by a used König which we had torn down and rebuilt to the same specs as the first, and on the shakedown flights we did have two forced landings. Fortunately they were never close to life-threatening. The problem was narrowed down to a plugged filter leading to the float needled in the tiny Bing carburetor. Once the problem was solved, there was never any hesitation from that engine.

The first day of goose-hunting season coincided with our first cross-country flight. We had chosen Herb Cunningham's strip, about thirty-five miles north of home base, as our destination. This would be the first time the birds were landed at a new location and the longest flight to date. The route had to skirt Lake Scugog and the marsh area to the north of our field. The day looked great, only a few high cirrus clouds and virtually no wind. This also would be the first shakedown trip of our ground crew and vehicles. A great deal was riding on this hop. We had no difficulty getting the eighteen birds that we now had flying with us off the ground and formed up on a steady fifty-foot-per-minute climb to the northwest. By the time we were abreast of Port Perry, we were seven hundred feet above ground. The air was rock steady and everything was working like clockwork. I flew the slow wing. Joe cruised high behind and to my right. One bird, 007, would often fly right beside the cockpit, so close I could reach out and let its wingtips brush my fingers. It could be likened to a friendly puppy anxious to stay by its master on a morning walk.

Once we arrived over the Cunningham field it was an hour-long drama of difficult flying to get the geese to land at their first "foreign" location. They were extremely cautious and made numerous low passes, lower and lower over the field, before their final touchdown. We then discovered that we had one more bird than we had taken off with; a wild goose had joined the group en route. We camped out in the migration project's motor home, guests of the Cunningham family while we were weathered in for three days. The stranger goose did not know what kind of a party it had crashed. To the other geese it was outside the family and was ostracized for the weekend. On the fourth day we flew home, birds in tow, without a hitch. The strange goose formed up and when we passed the point where it had joined, it dropped out. I watched it drop away to a small speck, landing in a farmer's pond.

A second cross-country flight was undertaken the next week to George Volks' field fifty miles to the west. It was below freezing on take-off and we had to defrost the wings. The birds were distracted by every gravel pit we flew over. The duration of this flight was more than two hours. At Volks' field we were again weathered in for several days.

Since May we had lived and breathed gas fumes and goose droppings, grown hoarse with attempted goose calls, hiked scores of miles, frozen our fingers and never ever gotten enough sleep, but by mid-September we were almost convinced that the birds were following well enough and the Airlie project would become reality. By mid-October we were ready to go, and the date chosen by Met Tech as being good for the lake crossing was Monday, October 18. We were in almost a military mode, the total scenario for crossing the lake worked out and carefully planned. Five aircraft, two rescue boats, and five land vehicles were involved, plus eighteen personnel, mostly volunteers. We held a brief meeting to ensure that our plans were fully understood, just about the time that Met Tech informed us that on October 18 we would face adverse winds, and should stand down one day.

CHAPTER ELEVEN

The Flight South

Approaching take-off time, now changed to October 19, the most important part, aside from the actual flight was, as before, ensuring the safety and effectiveness of all eighteen people involved and our equipment. In addition to two photo planes, there would be three aircraft dubbed Echo 1, Goose Leader (me); Echo 2, Goose Chaser (Joe Duff); and my old friend from twenty years earlier, Clark Muirhead, flying an amphibian Lake Buccaneer, code-named Papa Zulu November. Clark's role would be to range over what we saw as our principal potential obstacle, Lake Ontario, monitor our lake crossing, and perform or direct rescue operations if required — that is, if the ultralights flown by Joe and me ran into trouble, or if the geese decided to land in the water.

Then there were our ground vehicles: Groundhog 1, a Bendix motor home pulling the trike trailer carrying spare wings and the Cosmos two-place; Groundhog 2, my GMC truck pulling a trailer containing goose pens (one for each time we landed and another to leapfrog ahead for the next landing), pools, dog kennels, generator, water tank, hoses, electric fencer and motorcycles; and finally Groundhog 3, Don Lounsbury's Suburban with trailer.

In addition, two marine units dubbed Rescue 1, a twenty-one-foot vessel captained by Jim Bishop, and its backup, Rescue 2, would be on hand in event of lake landings that might be O.K. for the geese but not for Joe and me, landings in icy water that we fervently hoped would never take place.

While awaiting Met Tech's final go-ahead weather report, at our final briefing questions were asked and answered, and sometimes plans altered to make sure that all the activities of the various units were synchronized. When we took off with the geese, the marine units had to be already in place on the lake. The land units, leaving a day before the main flight, had to be ready to meet us on the U.S. side of the lake, and have pens up and ready for the geese when we got there.

Several pairs of headlights pierce through the trees into our kitchen. It is 6:00 A.M. and I have not slept since about 2:30. For weeks all kinds of responsibilities have been running through my mind. Will we have a repeat of the high-speed goose chases we sometimes experience? Will the geese get tired halfway across the lake and land in the water? Will we have a head wind and spend two tiring hours crossing the lake?

I remember that a Toronto radio station will be calling me at 6:15. I call Met Tech and they reassure me that the only thing that we might have to contend with is ground fog. Everything looks like go. I take a slug of tea from the goose mug and call Murray Cooper, who is in charge of the rescue boats. After several rings he answers with a hoarse voice and I tell him it is his wake-up call.

The kitchen seems to fill up with people and camera equipment. A camera crew from ABC-TV is hustling about, few saying anything. The phone rings. It's the radio station. They keep me on the line listening to a commercial for what seems forever, while I wait for the announcer. The questions somehow seem inane, but I politely answer them and sign off.

It's time to go to the airstrip but it is still pitch black. There is an overcast sky, no fog. Outside, Murray Cooper has arrived with the volunteer boat crews and picks up my son Aaron. The boat crews seem in a jovial mood. I have trouble being as jovial as they are. I am anxious for them to get going, because they should be at Newcastle harbour by 7:00 A.M.

Joe is already at the strip, his aircraft wheeled out, lit up by the headlights of his Jeep. I park my Pontiac to illuminate my craft and notice Iain Mellows uncovering his photo craft, Kodak 2.

It is still very dark. I walk over to the goose pen and say a few words to the geese, which are illuminated by the lights of the news van. Joe asks the news people to move the lights away. We are being extra cautious, nervous about upsetting the geese on this special day.

Back in the hangar under the Pontiac headlights, I start suiting up. The ABC crew wants to film it, but it is too dark. They ask me to wait five minutes while they set up a light. Reluctantly, I agree and check out the aircraft. The truck inner tube, a safety measure, in place, stuffed in front of the seat. I turn on the GPS (global positioning system) navigation unit and watch it find the required satellites.

The camera crew is ready. I fumble my floater suit on, attach the smoke flare to my waist, pull on the moon boots, fur collar, zip on the bib (for some strange reason I think about Neil Armstrong and Buzz Aldrin putting on space suits inside the lunar module before stepping down on the moon), glasses on, Michael Jackson gloves, and the bright orange inflatable life vest.

Colin Kemp is on the cellular phone. He has reached Clark Muirhead (Papa Zulu) in the amphibious Lake Buccaneer, waiting at Oshawa Airport; he is ready. Jerry Cochrane calls in from the boats stationed at Newcastle Harbour, reports there are light winds and a slight swell, visibility good. The boats, Rescue 1 and 2 are ready. Everything looks go. Iain Mellows is suited up in Kodak 2. He wears a crazy face mask that makes him look like a blue jack-o'-lantern, the crotch of his flight suit wildly patched with bright yellow gaffer tape.

The ABC camera crew is asking me to hold off launch until they have more light. I consult with Joe and again reluctantly agree to a five-minute hold.

I have a tough time holding back my impatience. The five minutes drag by. Son Geordie shows up and Richard VanHeuvelen is standing by at the pen, ready for the release. The aircraft engines are burbling away at idle. Dawn is retarded by incoming overcast.

I glimpse a few silhouettes on the ridge against the brightening sky, our neighbours gathered to watch the launch, with several curious cows bunched around them.

I am seated in my cockpit now. Richard hands me my mitts and camera. I look over at Joe and do a radio check.

Joe responds. I wave off Iain in the Kodak 2 and wait while he disappears over the ridge to the east. Richard is on the hand-held radio asking if I'm ready. He and Geordie are at the pen. It is time. I give them the thumbs up.

Slowly at first the geese emerge from their heaven and then at once they take wing. Full throttle, my craft, Goose Leader, heavy with fuel and safety gear, trundles down the strip and is in the air. The birds are off to my left. I veer toward them, just clearing the wind sock. They keep veering left over the forest and then over the field to the northeast.

Joe is airborne behind and to my left. He comes up on their flank. They loosely form up on him. He swings a wide arc to a southerly heading at about three hundred feet and passes the spectators gathered with the cows on the ridge. I parallel him to the east. The geese are between us now on a southerly heading. Joe crowds them father east and they gradually start to form up on his wingtip. We talk back and forth. "Echo 1 to Echo 2," I say. "Roger that," Joe comes back when he hears my voice.

Our track is a little too far west. We want to work farther east because a newspaper photographer has set up to get a picture near the village of Enniskillen. We miss the rendezvous by about half a mile. It is not a warm sky, full grey, and we have a slight head wind. The GPS is showing twenty-two miles per hour ground speed. The dark green of the great pine ridge has given way to the patchwork of southern Ontario farmland.

We grind southeasterly toward our Newcastle turning point on the lake. My earphones start to crackle. "Echo 1 this is Papa Zulu November over the lake at Newcastle. Over."

"Papa Zulu, this is Echo 1. We read you. Over." Papa Zulu relays the message to the rescue boats and they proceed out from the harbour.

The geese have broken away from Joe and are now flying midway between us in a ragged line. I call to them and gradually they move to form up on my wings. Several climb over the leading edge of the left wing, forcing me to push hard to the right to maintain course. The wing pulses with their wing beat. Eventually they settle out and string off the left tip.

The shore of Lake Ontario seems to take an age to arrive. We are about one thousand feet above ground. The air is stable. Passing over Highway 401, the geese slip away and join up with Joe.

There is what appears to be a low-line squall about five miles offshore. It looks black and ominous. Papa Zulu, reading my thoughts, reports that he is inbound from a ten-mile pass out over the lake and while there is a little turbulence at the cloud line, the situation beyond is smooth.

Aside from being ominous, the cloud formations are beautiful. Beyond, there are small patches of sunlight filtering through the overcast like ragged orange spotlights on a horizon-to-horizon azure stage. After what seems an hour but is in truth only a few minutes, we cross the shoreline. The geese break formation! Both Joe and I are silent. This has always been a concern: Will this huge body of water affect the birds? Will they decide to go down to land or turn back?

Our worries prove groundless. They move over and

form again on my wing. We now have a new heading and our ground speed has picked up to thirty-five miles an hour. Our fears of crossing the lake diminish in unison with the miles covered. The lead boat is able to stay right beneath us the whole route, while Al Griffen and Clark Muirhead in Papa Zulu orbiting about us, keeping a good distance.

On nearing the U.S. shoreline, we spot Clive Beddall piloting one photo plane, the two-place Cosmos. It's a thrill to know we've crossed the water. The birds are like a string of beads off my right wing-tip. We swing east now at about seven hundred feet. There is a slight haze and the smell of wood smoke from a farmhouse chimney welcomes us warmly to the United States of America. Our first destination, Frenchy Downey's field, is not hard to find. A hundred people are waiting at the north end of the narrow grass strip: newspaper people, TV crews, neighbours. I fly straight in and land at a remote part of the airstrip, assuming the birds will follow right in. It is not so. I land and ask the crowd to stay at the other end. Joe stays airborne and rounds up the birds. They are out of sight for a short period. We hear gunshots from the same direction — a worry, with the hunting season on — but the geese all appear again and after about ten near-landings they fly right by me and land among the crowd in front of the TV cameras. Everyone cheers!

The geese are penned. Carla Hengerer of U.S. Immigration kindly checks us in to the United States.

On October 20, ground fog and winds delay the next leg for two days. The local chapter of Experimental Aircraft Association invites us to put our aircraft in one of their hangars. We welcome the offer and ferry the craft a mile or so to the Gaines Valley airstrip. The next evening we attend their monthly meeting. Members have rolled their aircraft out of the hangar and put ours in — wonderful people.

Saturday, October 23, 7:30 A.M., we take off again. The birds form well. We set a southeast course to Carl Perry's field at Tuscarora, New York. A tailwind puts us ahead of schedule, so we overfly Perry's and carry on to land at Jalamtra field near Bath, New York.

I am on the phone to Met Tech. They advise us to carry on. The geese have already flown 2.2 hours. We take off again at 4:15 P.M. Winds are light but the air is still a little bumpy. We cross into Pennsylvania with a slight tailwind. Now we are in the mountains and have to climb to 3,700 feet above sea level to clear. The Apollo GPS, working well, takes us to Jim Finks' field in Trout Run, north of Williamsport, Pennsylvania. This airfield, called Finkhaven, is in the bottom of a valley with 2,000-foot mountains steeply rising on either side.

Joe says, "It's like landing in the bottom of a glass."

It takes twenty minutes to descend to field level with the geese. We land at 6:30 as darkness closes the day. The overnight pen is erected by truck headlights. We all go out for dinner and beer at the local bar and learn that the Toronto Blue Jays have won the World Series. It is a big day for Canadian birds in Pennsylvania.

Sunday, October 24. The weather report is iffy because we are in mountains. It is a clear dawn, no wind. The crew is slow to get started because of last night's celebration. Not so with the geese. They have the migration itch and are ready to fly.

The winds are still calm as we depart at 10:45 A.M. The birds seem a little disoriented within the valley. Joe and I fly several circuits before they form up on us. We set course at 180 degrees and fly for about one hour. We cross the Susquehanna to the west of Williamsport. I note the ground speed is down to eighteen miles an hour and the air is becoming turbulent. We carry on over the first ridge of mountains. A big bowl to the west causes turbulence from southwesterly airflow. We get severely bounced. I am above, Joe is with the birds. I look down and see our string of birds, led by the moth shape of Joe's Cosmos, creeping sideways across the terrain. Our ground speed is down to fourteen miles an hour. We spy a clearing in a valley with a track down the middle, and after several circuits I land. The geese land with little problem and Joe follows in. Our cellular phones do not work. I find a phone at a nearby house and catch the ground crew still at Finkhaven. We have an unscheduled camp-over much to the surprise of the Miller family whose field we have dropped onto.

October 25. We take off at frosty first light, climb eastward for five miles before we have enough altitude to clear the first ridge of mountains. Our climb continues to cruise at 3,500 feet above sea level to clear subsequent east-west mountain ridges. Ground speed reading is fifty-four miles an hour. There is fabulous visibility; smoke from many wood stoves rises straight up. The Susquehanna, shrouded in early morning mist, twists southward off to the east. We skirt to the west of Harrisburg control zone and decide to cancel a scheduled landing (Lester Yost's field at Sherman's Dale) and keep going.

When the air is perfect like this, without problems, my mind sometimes moves into another mode, one of deep reflection. Once we have reached cruising altitude and the formation of aircraft and birds has stabilized in relative calm, I can watch the bird flight in detail. Most of the geese will fly in a line, surfing the pressure wave of the invisible wake that originates at the wingtip and extends back from the aircraft like a boat wake. One or two will fly within arm's length of the cockpit, affording me an armchair view of the bird in its own element.

Every part of the creature is a wonder of aerodynamic efficiency, a poetry of movement that extends from the head, the stable "centre" of the bird, to the fluid dynamic action of every other part of the body. The nervous system controls the neck muscles which constantly flex just the right amount, maintaining the head stable and balanced while the body undulates rhythmically up and down in reaction to wing strokes. The term "flapping" to describe the motion of the wings seems a crude simplification. The wings and feathers all move in a complex symphony, each feather playing its delicate role in the concert of flight. The smallest, innermost feathers make the subtlest of movements while at the extremities, doing the most work are the primary feathers. These long slender feathers are a perfection in design and action, each having its own unique shape, twist, and resilience. The opposite wing is a perfect mirror image in shape and motion. The birds' sensitivity to the air movements is another wonder. They quickly sense the perfect spot to fly in relation to the aircraft wing, to optimize their lift and thus get the easiest flight. They often will form in a line along the leading edge of the wing, taking advantage of the pressure wave over the wing, even though they often spoil the lift of the wing. If they are all on one side, it becomes almost impossible to maintain the aircraft in level flight. The action of the wing beats is transferred through the aircraft wing, and becomes quite a pronounced pulsing which I can feel through my control bar.

The geese also sense temperature and pressure; they know when the flying is easy. If they fly in warmer weather and begin to overheat, they lower their feet, which affords better cooling. They have many manoeuvres which control their speed and altitude. I have watched as they catch up to the aircraft and then need to slow down to match the aircraft cruise speed. To do so they arch their necks, which increases drag and slows them down. Also, all pilots know that to lose altitude you cannot just point the nose of an aircraft down without gaining speed. The geese use the technique of whiffling: they just pull in one wing, flip sideways or even go upside down, and in an instant they are down ten or fifteen feet and are still going at the same speed. When they do this in a flock, it is a random event started by one, then quickly copied by the rest. One whiffle might follow another until they are tumbling randomly like fluttering leaves, until again they form up stabilized at a lower altitude.

While I am watching and reflecting that day, we all carry on in that perfect air to land south of Gettysburg, Pennsylvania, at Brown's farm.

No one is home at Brown's farm. The ground crew takes two hours to catch us, then they miss the driveway. I jump up and wave frantically at them and inadvertently scare the geese into the air. I almost go into shock, but my fears melt when in ten minutes they return and land at my feet.

October 25, 3:30 P.M., 70 degrees Fahrenheit, still clear, beautiful "designer" weather. We get airborne for the last leg into Airlie. It really is too warm for the geese, but the forecast is for bad weather the next day. The geese pant the whole two and one-half hours and will not climb. We have to maintain low altitude, five hundred feet over beautiful Virginia rolling countryside and fabulous homes. Airborne spiders that ride the wind on long silk threads collect on our craft and their streaming webs shimmer in the afternoon sun. The Apollo GPS brings us to Airlie at 5:30 P.M. Dr. Sladen awaits and again there are the film crews and spectators. The afternoon light is perfect, the air is absolutely still, the whole scene is surreal, and the geese need water badly. Our ground crew arrive simultaneously and set up the pens.

On the way into Airlie I had been remembering a lot. There was a point in midsummer, the first few weeks of attempting to get the birds to form up with us, when we were ready to fold up *our* wings. It seemed we were always on the uphill side of the curve. It was the change in wings and a new approach that gave us enough hope to carry on, and gradually the project turned around. It certainly was a glory day when we landed in that brightly lit Virginia meadow at Airlie with all eighteen geese. The significance of it all began to sink in — along with occasional thoughts that this was only the first phase of our migration project. Still ahead was the key test: Would they complete the migration by retracing their flight the following spring, as wild migrants do?

CHAPTER TWELVE

Airlie Winter

Resting at Airlie, we had a chance to learn and enjoy. Airlie Center was once dubbed "an island of thought" because it offered a complete sanctuary to groups who wished to get away from the everyday life of their urban environment, enabling them to better plan for the future — whether that be to make more money or to save the world. Located outside of Washington, D.C., it was assembled in 1959 from a group of near-abandoned farms by the late Dr. Murdock Head, who had the vision to create a conference and research centre in a quiet, comfortable, and protected atmosphere. Airlie does just that. It is a glorious setting for people to have meetings, become re-energized, and enjoy nature at its best. Picture this: a Georgian Revival mansion and restored cottages tucked in beautiful countryside, each building a poetic name such as The Manor House, Groom's Cottage, Raven's Hollow, Clifton House, and the like. The 3,000-acre property encompasses forested hills, stone walls, and boxwood hedges, and is dotted with ponds, waterfalls, and marshes for the many thousands of birds that find their way to Airlie each year, either to spend the winter or to stop off and be re-energized before continuing their migration. They say that George Washington used to travel the roads at Airlie. Today, when you walk those same roads, you almost feel as if you were walking in his footsteps.

Dr. Sladen is truly the birdman of Airlie. His wife, Jocelyn, could be considered the plant lady, for she is most knowledgeable about the botany of the area. They live at Clifton Farm, a marvellous old house on 800 acres set off from the rest of Airlie, down a twisty road through the woods and past two large ponds that are home to both resident and migratory waterfowl.

The Sladens enjoy wild plant life. They promote the growth of natural habitat and fence rows of native plant life. Clifton Farm has become wild and natural, and is a true contrast to much of Virginia's manicured lawns and trimmed hedges.

Two blinds front on the larger pond at Clifton, and the social event of the day is to sip sherry with the Sladens in one of the blinds at sunset to observe the swans in beautiful proximity.

After our group of eighteen geese arrived at Airlie, Dr. Sladen soon dubbed them the Ultrageese and neckbanded them with what looked like small soup cans! The birds didn't like this a lot, but it enabled us to identify the individuals in the air and on the ground quite readily. Practical but ugly!

The birds were penned at the start, but we flew them daily to help them learn the area and accept it as their winter home. Then gradually they were weaned from us and by January 1, when the goose-hunting season closed in Virginia, they were allowed to go free.

That was the coldest winter in forty years. The ponds at Airlie froze; electrically driven bubblers kept open some spots so that the birds had access to water. Two of our birds died over the winter. One must have been frightened down under the ice, and a few weeks later was found drowned. The other had to be put down because of an injury. Over the winter Joe and I rotated going down to Airlie to take the birds for flights whenever weather allowed, wanting to have them in shape for the spring, when we would see if they would return to Ontario on their own when other flocks of migrant Canadas left Virginia.

On April 1, they were still at Airlie. We figured that if they did not migrate back to Canada on their own, We'd round them up, have them form up on the ultralights as they had the previous fall, and fly them back to Ontario. The last of the migratory waterfowl had departed Airlie two weeks earlier, meaning that our geese might need a bump start to get them flying north. But the next day, where were they? Just gone! Had they headed home? That was the big question. We did not really worry. They had been free geese since the first of the year and had already gone off on their own, little jaunts lasting a few days at a time. We figured we would just wait till they showed up again, then get them airborne and start our northerly navigation.

A week went by. Not a feather of them to be seen. We got the two Echos, Goose Leader and Goose Chaser, airborne and flew around the vicinity checking out the various ponds where they had been seen. No Ultrageese.

Jack Weber, an old friend and flying buddy, had accompanied us on this trip to drive the motor home back should we fly. Joe, Jack and I became a little worried. We wanted to get moving. Waiting for the geese to appear was like waiting for Godot. In Airlie on April 14 — still no geese. We decided to pack up the planes, head north, and look for our geese at our previous stops en route.

Gettysburg: no geese. Trout Run: no geese. New York State: nowhere even to start looking.

By the time we got to Cohocton, New York (the speed trap capital of the United States), we were depressed. One more stop was planned at Frenchy Downey's, the last likely landing spot before Lake Ontario. I cranked up the cellular phone and punched in Frenchy's number. His wife, Dorothy, answered. I told her of our planned visit. For a moment there was a silence. Then she asked, "Did you talk to your wife today?" I hadn't. She told me Dr. Sladen had called an hour previously and that he'd had a phone call from Paula.

On her morning walk Paula had been greeted by ten of the Ultrageese at home on the Purple Hill runway! They'd made it home! Waaa hoo! We just had to pull the motor home off the road and jump up and down and yell for a few minutes. The geese had made it back! What a day! Richard had recorded the neckband numbers. A couple of days later two more showed up — twelve now out of sixteen.

Paula's own account later told the story of the big day. "I had gone out twice that morning. I had got up early.

Geordie had to be in Blackstock at 6:45 for a driving lesson, so I got up, took him out at 6:30 and I didn't hear any geese, I did listen, but I didn't particularly look down in the field. I got back around 7:05 or so and half an hour later I left to go for a walk with our neighbours, Lynn Wonnacott, Bill's secretary, and Joan Graham. When we were walking in the field next to our field with their dogs, Lynn looked over at our runway and said, 'Gee, that looks like geese over there.' Of course we all looked. We could see them! We yelled and made a beeline for the runway. I went up to the geese. Lynn and Joan stayed back and held their dogs because we didn't want to scare the geese into the air, but when I got close I could see their neckbands and knew they were the right geese, twelve of them, or I guess ten that came back that first time — ten or eleven cause they had one or two wild ones with them. I went toward them and they all came over to me. 'Yayyy!' I was yelling. 'You guys made it! Oh, look at you, you flew all the way back!' You know, congratulating them. It was really exciting, just talking to those birds and bending down, and they didn't know me really well but they knew me enough that they came over, like they do, with their heads bobbing up and down — 'Oh, look at you guys!' They were so proud of themselves, you could tell. They were just doing their little low sounds, like cackles. Feeling really contented and proud and happy… We got Richard to come down and we wrote down all the numbers on the neckbands. Then we put the birds in the pen. We didn't want to leave them just free. We didn't want them to take off before they made the connection with Bill. I even thought maybe they might be thinking, 'Where's Dad to say congratulations?" In the pen, even with fresh food and water, they were really unhappy, as if here they had done this fantastic feat of flying for two weeks to make it all the way home and what do they get — the pen!

The wild ones just took off. One of our geese took off with them. We got the rest in the pen. That night when I was up at the house the wild ones flew around and came back. We could see them outside the pen. All night long it was yap, yap, yap. They were just talking away, three outside the cage free and the rest of them in the pen, really upset. Bill made it home the next day, so we only penned them up that one night."

That was the scene I came home to. How did they do it? What route had they followed? Many questions raced through our joyful minds. We could not wipe the grins from our faces as we motored those last miles to Purple Hill. Our reunion with the geese was just plain fun. We went to the pen and let them loose and they made straight for the pond for a quick dip. For two days they hung around and then they flew off to join other wild flocks. All the local papers ran stories about their return. I cannot count the number of phone calls we received, either congratulations for the geese or from people who had seen a goose with a neckband, thinking it might be one of the Ultrageese.

As I wondered how they had remembered the way, I thought back to my knowledge of goose intelligence. The Native American perception of wildlife is that there are many spirits in nature, a collective consciousness particular to each species. This consciousness could be thought of as the god or goddess of each species — the beaver god, the wolf god, the goose god. It is a way of understanding otherwise unexplainable mass behaviour of different species. Scientists usually cannot accept the perception of a deity, a collective conscious spirit for each particular species, because there is no way of proving it one way or the other. Being an artist rather than a scientist, I have the freedom to believe things without having to prove them. I believe there is a collective conscious spirit, a hierarchical array of deities within the realm of nature. I mean who is Mother Nature other than the overall consciousness of all life? Considering the myriad of fantastic life-forms on Planet Earth, I believe the process we call evolution is a creative continuum, definitely not some accidental process loosely called survival of the fittest. Although accident does play a role in evolution, the process is far too creatively organized to be sloughed off to accidental development alone. As for science, it is an excellent tool for helping us understand the details and workings of this marvellous creative continuum. There are constant and wondrous revelations through science.

While at Airlie during one of my winter visits to fly with the geese, I had had a dream for my retirement years of getting a house somewhere in southern climes to avoid the rigours of cold, dark snow-blowing days. The dream gave rise to comparing how we humans migrate in comparison to the birds. So many older Canadians trek south in winter to escape having to struggle with the aching realities of Canadian winter. They stay in Canada from the beginning of April to the end of September — six months. Then they drive or fly south to Florida or Arizona or Mexico for six months. It is not only ice they escape, it is those long winter nights of darkness. Birds of course have been doing it for millions of years. We use the term "birdbrain" derogatorily. Are birds stupid? Let's think about it for a while.

Perhaps it is geese, the smartest of these birdbrains, that migrate the farthest. They fly north in the spring. Taking their time, they move with the weather, optimizing both the spring and the daylight. Having the freedom of air, they pick quite suitable spots, five-star goose hotels, on their northward journey while enjoying the wonders of an ever-changing world in the process of its spring awakening. When they get to their northernmost summer habitat, they enjoy the longest of days. In remote pristine estuaries and wetlands, during the long golden twilight of the north, they do their courting displays and partake of their annual romance and lovemaking. As a team, couples raise their new flock of young and stay together as a family unit. In remote spots safe from predators they renew their feathers. Losing all the year-worn garb, they emerge renewed in fresh attire as the summer closes. The days grow shorter and they gather to start their journey south. They don't have all the encumbrances of clothing to take with them or a house to close down. Their comfort is self-contained, the only upkeep their feathers, which they preen regularly. They have no alarm clock making them report for work each day. They don't have to deal with automobiles, airports, traffic jams, monthly bills, income tax, garbage, dirty laundry, border crossings, and on and on. They know when to travel with the weather. When the systems are moving the way they want to go, they hitch a ride on a moving air mass and boogie southward a few miles or a few hundred miles at a time. Flying together in great Vs, they enjoy the camaraderie of their kind and fly with far greater efficiency than if they travelled alone. The older generations direct the flight to comfort, food, and security. Stopping at their favourite spots, they renew old friendships and make new acquaintances. They munch on an abundance of the harvest in the stretched-out autumn of their southward journey. To the north the days grow shorter, shutting down plants and resident animals for the dark, dormant winter, but the geese have escaped the harsh wind-driven sleet and those long dark nights. Stupid? I wonder....

There are social strata among a flock of geese not too unlike the human equivalent. We know it simply as a pecking order. And there is a full spectrum of characters within each flock. The gregarious, the surly, the bully, the sweet and friendly, the oddball, the timid, the adventuresome — they are all there, creating a healthy social blend. It is deeply meaningful to me that geese, or I should say birds in general, have not gone through any major evolutionary changes biologically for thirty-five to forty million years — a stretch of time which predates by a factor of ten the beginnings of our human species.

This comparison gives rise to speculation about their origins. Could it be that birds had origins similar to us? That once they had large brains as we have? I mean, perhaps they started out evolving first into bipedal humanoids paralleling our early development. They could have evolved out of dinosaurs, walked erect, had arms, hands and large brains. In those millions and millions of years they may have evolved to a highly complex intelligent culture. They may have explored the world's developed technology as we are in the process of doing. Maybe their evolution at this point did not parallel our thinking. Perhaps they thought it out well enough ahead to realize that the highly developed technology that currently is part and parcel of our culture and comfort is all too crude and inefficient, that the creation and maintenance required by this technology is wasteful and not at all in keeping with the other evolved natural systems of the planet.

Perhaps they foresaw the problems that would ensue with overpopulation and that the answer was to control their own biological development. In short, they reached the point where they found they could engineer their own genetics and that by taking control of their own biological development they could control their environment and ensure their survival. They found this bioengineering a much more refined technology than the crude mechanical systems they had begun to develop (is this starting to sound like science fiction?) Is it possible that they thought about the whole interrelated web of life on this planet and how they might most successfully and efficiently enjoy the wonders of it, while socially and spiritually developing among their peer group?

Perhaps they discovered that the best way to optimize their social lives while enjoying the beauties of the planet with its changing seasons, weather patterns, and the great variations in fresh food supplies was to evolve into a creature that could navigate in the boundless ocean of air that surrounds the planet. Perhaps they realized that a better system would be to develop a minibrain connected to a bird internet, a brain like a personal computer connected to all other personal computers. This allowed them access to the mainframe, which was actually the whole species interlinked, and this group consciousness, this telepathic system interconnecting all members of the species, allowed them global consciousness of world weather systems.

And perhaps they also were able to design their body, one that could be sustained efficiently on the simplest of plants that occurred in greatest abundance. They understood that the safest and greatest thing was to have the freedom to be mobile on or in any of the mediums on the planet's surface. They could swim, dive, fly, or walk, depending on the need. Because they had this internet-chip telepathic brain as their main system of communication, their relationship with their fellow birds could be intricate and direct, with only the use of simple verbalization and body language as a vestigial secondary communication system to relate mostly to other species that were not tuned in to their species' specific internet. What a life they could lead with their families, travelling across the planet with the seasonal changes!

This is probably all too fantastic a concept. For the world of the goose, wildly idyllic in some ways, is subject to many hazards. Many have ceased to migrate for they have lost the knowledge of the migration route. They can stand extreme cold as long as there is open water and somewhere to feed. Therefore they hang out at city parks and near the warm-water discharges of power stations, and become winter panhandlers. Predation is the cause of mortality of most young geese; owls, hawks, and eagles attack from the air with impunity. Owls will even grab a full-sized goose. On the ground, foxes, coyotes, wolves, weasels, mink, otter, wolverine, raccoon, and the various wild cats all dine on goose or goose eggs. In the water, goslings disappear into the jaws of snapping turtles and large fish. And of course there is the wiliest of all predators, that upstart bipedal creature called man.

Prior to controls on hunting there was virtually industrialized mass murder of waterfowl. Cannon-size punt guns that blasted flocks of waterfowl in the water could kill a hundred geese with one shot. They were common for many years in the 1800s. All of these factors decimated the Canada goose population by the 1940s. They could have become extinct had we humans not put controls on our hunting habits.

Now the goose has come back in strength — such strength, in fact, that in many places Canada geese are now considered a nuisance. Larry Hindman, a goose biologist with the state of Maryland, estimates there may be half a million so-called "resident geese" now populating the Atlantic flyway between North Carolina and eastern Canada. These are geese that do not migrate. They are the product of decades of policies by Canadian and U.S. wildlife managers that allowed private individuals to raise geese on the theory, the more the better. They have adapted all too well to a suburbanizing countryside, finding abundant food on golf courses and municipal parks. In recent years, aided by a series of warm winters, their numbers have reached a critical mass, and they are now on the verge of exceeding the number of migrating geese in the flyway, with no end to their increase in sight. They are causing problems at airports, running into planes, and rendering sections of golf courses and parks nearly unwalkable, as well as polluting ponds and streams with their waste. "My prediction is they are going to be a bigger nuisance than the whitetail deer that have adapted so well to suburban, and even urban situations," Hindman says.

Some biologists, like Hindman, think the resident geese are overwhelmingly from crosses of Canada goose subspecies that never migrated. Dr. Sladen and others share the belief that a good many are just geese who have never learned to migrate. He cites the example of five "controls", geese that were taken by vehicle to Airlie at the same time we were flying down with the Ultrageese. These five, who never learned the route, did not migrate back with the Ultrageese when released in the spring. Along with Dr. Sladen, I like to think that just maybe we can train numbers of these resident nuisance to recapture the ancient migration urge.

One thing we know for sure. The Canada goose can always surprise us with its responses to changes that humankind is constantly making to the planet. A wonderful example is what has happened in just the last four or five decades — the geese responding to something that might seem very unconnected to such wild creatures — the rise of a huge domestic poultry industry. On the Delmarva Peninsula, where Delaware and parts of Maryland and Virginia form the eastern side of Chesapeake Bay, some 360 million chickens a year are raised for meat. In the 1930s and early 1940s, when this industry was just beginning to grow, Canada geese were relatively scarce in the Chesapeake region. North Carolina was the goose capital of the East Coast then.

But as poultry numbers rose, so did the acreage in grain to feed them. Harvest technology was also changing to mechanical combines, which were efficient, but left a good deal more grain lying in the fields. The geese, figuring why fly to North Carolina if you can be fat and happy in Maryland, began to change their migration patterns. By the 1980s, there were close to three quarters of a million Canada geese wintering around the Chesapeake (poor breeding weather in Canada for several years has since reduced migratory populations in the whole flyway).

Unknown to scientists and game managers, there was an ecological factor at work here too. Polluted runoff into the Chesapeake — fertilizers from grain farms and manure from chicken operations, was building up. In the 1970s this contributed to huge die-offs of the bay's once-abundant beds of underwater grass — prime goose food. This accelerated the movement of geese into the farm fields and grassy areas on land. Populations of waterfowl that proved less able to switch, like canvasback and red-head ducks, crashed baywide, and have never rebounded despite conservation efforts. Dr. Sladen often recounts this story to make the point that an early-warning signal of the demise of the bay grasses was the mass movement of the tundra swans from feeding in the shallows to invading fields to forage. Having ignored such signals, Maryland and Virginia now are spending hundreds of millions of dollars to reduce pollution and restore the underwater grasses, which are also vital habitat for everything from blue crabs to dozens of fish.

The migratory ways of birds have always been a great enigma to us land-bound creatures. Exactly how did our geese, who returned to Ontario on their own in that spring of 1994, learn the route back? What route did they actually follow? Did they go from airstrip to airstrip, remembering the features of the terrain, or did they tune into the sun and stars or earth's magnetic field — or a bit of all these and perhaps more? We would like to know those things. The scientific aspect of our project is most fascinating. After this first success we vowed to repeat the experiment, do it better, and learn more.

CHAPTER THIRTEEN

The Second Flight South

In the final count, thirteen of the 1993 Ultrageese returned to Ontario on their own in the spring of 1994, a group we felt was too small to form the nucleus of a migratory flock. The thirteen had dispersed among local wild flocks (one was seen with eighteen wild friends on July 21, fifty miles from Purple Hill). It was anyone's guess whether they would regroup and fly back to Virginia in the fall of 1994 or whether they would remain with the resident non-migratory flocks of southern Ontario. The experiment needed to be repeated with a new, larger flock and we needed to find better solutions to some problems we had encountered in 1993.

First, we rented a hundred-acre sod farm a few miles distant from our home, and the birds' home, at Purple Hill. The flat reaches of the sod farm would afford a greater degree of safety than flying out of my restricted strip at Purple Hill. Then we had the tremendous advantage of the involvement of the Airlie team of biologists. Gavin Shire came up from Virginia for a few weeks and oversaw the incubation of the eggs. After the goslings had hatched, Kirk Goolsby, one of Dr. Sladen's protégés, arrived and stayed. By 1994 I had imprinted a multitude of flocks over several years. Until Joe Duff came to assist me in 1993 I had always shouldered the major part required of the daily concentrated dedication to young geese, for periods of six or more months. Fortunately Joe was still inspired. Taking up residence at the sod farm, he worked daily with the young birds.

For the early imprinting of the birds, I had devised a wearable aircraft which resembled the real Goose Leader. It was half the size and had straps allowing it to hang on your shoulders. Gavin Shire dubbed it "Goose Toddler". There was no lack of laughter when it was put to use, particularly by Joe, who pictures himself as a top-gun pilot. With an embarrassed look on his face and mumbling to himself, he trotted up and down the field with engine noise from the built-in tape recorder at full chat, and a string of goslings in hot pursuit. It was a scene right out of Monty Python.

While Joe worked hard and suffered indignity to his image, most of the credit for the imprinting had to go to Kirk Goolsby and my son Geordie. They developed the most intimate relationship with this new flock. Kirk, except for the occasional day off, literally lived with the birds day and night. He tented by their pens at night. During the day he was never far from them. In fact I began to wonder if the Gary Larson cartoon entitled "When imprinting experiments go awry", showing a guy following a line of geese crossing a road, applied to Kirk.

To me it was really interesting how these two young guys, Kirk and Geordie, related to the birds. I had a great sense of fatherly pride watching how Geordie became involved. Kirk took a more scientific line, making notes at every movement and twitch the birds made. Geordie just related to each and every bird individually on a level that was inspiring. It was like watching new parents with thirty-eight babies while Joe and I played grandparents. All the birds were numbered and carried leg bands, but many of the birds were given nicknames by Geordie and Kirk.

There was Igor, number 38, the second gosling hatched that spring. He had a slightly misshapen head and walked with a faint limp and thus was named Igor after Frankenstein's assistant. He became a favourite and also for a time the most dominant of the geese.

Fleck Neck was so called for a strange growth of feathers growing from the back of his neck as if some of his breast feathers had been transplanted. This gave him the appearance of a stuffed animal with some of the filling sticking out. Like his mutant feathers he was a bit of a rebellious personality and would always be difficult to put back in the pen while the rest of the geese would walk in with ease. Yet when he began to fly he was a leader and could be easily identified in the air from his feathered outgrowth.

Number 25, Ringneck, was a different matter, a plucky female goose that survived an attack by a great horned owl. The owl had killed six of her siblings and had left a half-inch strip around her neck completely devoid of skin and feathers. For a long time we thought she would not survive but she did, thanks to some care from Dr. Mike Taylor, who for the second year was the project veterinarian. Also, Kirk did a bit of adept suturing which I am sure was the stitch in time that saved Ringneck.

Egghead was an interesting character who became Geordie's favourite. When hatching, she had trouble getting out of her eggshell and got stuck for a day or so. We had to break off the dried shell, leaving her with a big lump of yolk stuck to her head for almost a week. The other goslings took great pleasure in pecking away at her head. She did not gain weight for the longest time. We thought she would not make it to adulthood. Geordie claimed she was brighter than the rest of her group. While going on a walk in a line, one gosling after another would fall in one particular hole and have difficulty getting out. She would pause and walk around.

Crash, on the other hand, was not too bright. He flew into the power pole near the pen not once but twice, and another time hit power lines. He dropped out on practice flights several times during the summer and we had to go find him by road and truck him home. We thought he might have been partially blind.

There was also a protector goose, Barnstormer, with a low voice and very friendly disposition but requiring your undivided attention. When communicating with you he kept his head high, whereas the other geese communicated by lowering their heads and bobbing. Danny Brown, a five-year-old at our stop near Gettysburg on our route south, gave Barnstormer his nickname. He was not a fight instigator but was always in fights with other geese.

Freckles, also named Jeremiah, was the most reclusive goose and always hung back at the edge of whatever was occurring.

Christmas Goose, number 47, got his name because of his persistent waywardness in the air. I don't know how many times during our summer practice flights he would just break out of formation and head home, on occasion taking the rest of the flock with him. Several times he would turn around immediately after take-off and land back at the pen, confusing the whole issue. If any goose was going to be up for being Christmas dinner, he had the number-one ticket.

The list goes on. There were Spanky, Spot, Peckerhead, Peppy, Coffee, Sam (now, who would name a goose Sam?), O.J., Ogar, Eyes, Roman Nose, Homer, Bugler, Clunkhead, and then a whole bunch more that only were called by their numbers.

The sod farm with its great expanses of manicured turf was a joy to work on and fly from. Taxiing with the young birds became a daily routine. By mid-July they were flying and by mid-August they were airborne for half-hour stints. By October, thirty-eight geese were ready for the flight south.

Again we had in place the safety team to help us cross Lake Ontario, whose forty miles of open water remained probably the biggest worry on the whole migration route.

On Tuesday, October 11, 1994, a large atmospheric high had settled over our planned route. Everything looked perfect. In comparison to the previous year, Joe and I were relaxed. We even got some sleep. I got up several times during the night to check sky and winds. Prior to midnight the winds were blowing strongly out of the northwest, but by 5:00 A.M. everything had calmed down as predicted. By 7:15 A.M. we were suited and looking like yellow-collared black Michelin men with our inflatable Mae Wests, floater suits, layers of underwear, smoke bombs, and neck-worn emergency locator transmitters.

There was a tinge of frost on the ground. The aircraft gave us a little problem in starting. We eventually started the engines by hand flipping the propellers the old-fashioned way, warmed them up for a few minutes, then wedged ourselves into the fuel-laden craft and taxied out to the take-off strip. By radio, Joe and I debated whether we would remain airborne to pick up the birds or take them off from the short strip near the goose pens. With our fully laden craft, the take-off roll was so long that the airborne pickup was the only workable plan. By then the ground crew, Kirk and Geordie, were awaiting us at Frenchy Downey's field on the American side, and Richard was at the pens, with a faulty borrowed handheld radio. The communications were spotty but Richard understood our "go to it" call, released the geese, and the birds were airborne without a hitch. They were soon formed on Joe's wing and headed south over Highway 7 in the growing light of an absolutely cloudless sky. I hung back and watched the birds from a distance, enjoying the imagery of thirty-eight goose beads strung off Joe's right wing, flashing in the dawn sun while the dried corn and stubble fields of southern Ontario slid quietly beneath us. The line of birds undulated and swayed with each small wave within the atmosphere, resembling a streamer gracefully flowing in the breeze.

Crackle! My radio burps to life with a call from Papa Zulu; the familiar voice of Clark Muirhead is on cue as his cover aircraft orbits Newcastle Harbour ten miles ahead. Then Jim Bishop tunes in from the rescue boat.

All goes according to plan as we cross over Newcastle Harbour at about eight hundred feet. I see a number of well-wishers waving to us from dockside. The lake is brilliant blue and, as happened the previous year, we feel like tiny infants over its great expanse. Our two miniplanes with their following of naive five-and-a-half-month-old geese cruise over the water with no hitches. We speak frequently with the only rescue boat we see. It is the second boat and by mid-lake it is directly beneath us. Meanwhile Papa Zulu, the larger aircraft, cruises by in an elongated oval pattern. We begin to get a little concerned about the lead rescue boat, nowhere to be seen and with no one able to make radio contact. Papa Zulu scouts ahead to the American shore but still there is no sign. Rescue 1 attempts to make contact by cellular phone, to no avail.

About halfway over, Papa Zulu November reports there is a lone bird behind us. I ask if the bird is flying with its neck raised. It is. I had some idea that it was number 44, Crash. Several times during training Crash had dropped out and had to be picked up by truck. Just before he quit he would fly with his neck almost vertical, as if panicked. From the start we were not hopeful that he would make it. We had discussed the possibility of leaving him behind but had decided to let him come along. If he dropped out, we were going to leave him and not jeopardize the rest of the flock. In the past they didn't seem to notice his departures.

The birds have now lost a couple of hundred feet of altitude and we are down to five hundred feet off the water. By the time we reach landfall they are down to two hundred. Again it is a great relief to feel the comfort of open fields beneath our wings as we swing east and pick up a heading for Frenchy Downey's field. A little head wind slows our progress, but two hours and five minutes after take-off we are rolling down Frenchy's grass runway, all thirty-eight birds on hand, leaving us with no information about the reported lone bird. The geese all helicopter right into the crowd of about 150 spectators who have come to witness the avian invasion. I don't think they expected to experience the birds in such close and dynamic proximity. Children mingle with the geese, offering them tufts of grass, as flash and video cameras are overworked.

We enjoy our visit with Frenchy and his friends, a wonderful and much relieved occasion for all. His ninety-eight-year-old mother, who has been sitting on the porch awaiting our arrival for the past three days, is escorted out to greet the geese once they are secure within the portable pen. Kirk busily weighs the birds. The whole crew, geese included, are treated like royalty. Frenchy has just completed a 1930s-era home-built aircraft that he started in the early fifties, quite a wonderful machine powered by a 1930 model-A Ford engine. He fires it up and gives us a demonstration flight, a backward glimpse of a romantic era of aviation. That evening the Gaines Valley chapter of the Experimental Aircraft Association throws a potluck dinner in honour of our visit. Joe and I tell them stories of last year's flight.

Seven A.M., October 12. There is a good layer of frost on our wings. We work at scrubbing it off but the best job is done with Frenchy's garden hose. Again it is an air pickup. The geese initially take off to the west but are soon again connected to Joe's trusty right wing.

Our plan is to fly to Jalamtra field at Bath, passing by Carl Perry's. We cross the Erie Canal at about 150 feet, then divert to the west of a prison complex, with its surrounding razor wire sparkling ominously in the dawn sun. I see a group of prisoners stilled by our passage and consider the contrast in freedoms, all too apparent. By the time we reach the New York Freeway to the east of Batavia, our ground speed is down to twenty-four miles an hour and dropping. I do some calculations and at this rate it may take more than three hours to reach Bath. Joe and I confer and decide to alter course for Carl Perry's field to the south of Mount Morris. The ground crews, headed for Bath, need to be informed. We have lost radio contact; the only alternative is to attempt to use the cellular phone — a no-no from the air because it confuses the system. I drop down to 100 feet over open fields and am successful at raising Richard back home.

Carl Perry has one of the most serenely beautiful airstrips I have seen, with an adjacent pond that is soon home to the geese after we land. Carl has also been alerted to our arrival and is waiting. In a short time we are refuelled and airborne, but now it is 11:00 A.M. and the foothills to the Appalachians are getting higher. We skirt around the first, for the geese have not attained enough altitude to go over the top. Soon we are getting kicked about by the midday turbulence. Joe says, "Why were we so stupid as to take off at almost midday?" I cannot help but agree. The geese are having a hard time keeping up with us. Because the closer we fly to the ground the more turbulent the air is, we stay a few hundred feet up while the birds below take the same general path but close to treetop level. One of us hangs back until the lead plane outdistances the geese; then the other moves in to take up the lead. The turbulence gets somewhat frantic.

Joe says, "I'm landing." I agree. We find a big field on top of a lonely hill. Joe is first to land, while I maintain altitude until the geese surround him on the ground.

"One hell of a rough field," Joe radios to me, but I spy a side road running parallel to the field where he landed. There are no power lines to impede its use as an airstrip. It is also aligned into the wind, a gift. I am soon on the ground about a couple of hundred yards from where Joe and the geese are ensconced in the alfalfa.

A few minutes later a lone goose appears and lands beside me. It is Crash bringing up the rear. Where had he been?

Again I try to make the cellular phone work. Eventually I reach Don Lounsbury, who has already arrived at our destination, Jalamtra. I am in the process of telling him that I haven't a clue how they can find us by road when a well-worn pickup pulls up beside me. (I am standing in the middle of the road with the aircraft shoved off in the ditch.) Jim Wellington, in the truck, is the farmer whose field we have dropped on. Jim passes on the directions by road to the ground crew. About an hour later our whole column of vehicles appears framed in dust down the distant road. The geese are then secured within the portable pen, surrounded by half a dozen well-wishers who had led the ground crew the twenty-five miles from Jalamtra field.

We lunch in the hundred-year-old general store in the village of South Dansville. By five the winds have dropped, the geese are looking anxious, and we take off for Jalamtra. There we are welcomed by Herb and Shirley

Townsend, who have put on a large spread for us topped by grape pie, while the geese in their pen dine on their regular corn. In the morning we once again have frost on the wings to contend with. Our two craft, fully refuelled are at their maximum weight and, like the geese, we always preen our wings to make sure they are at their most efficient.

A day earlier we had met the Abbeys, a husband-and-wife teaching team from the Tuscarora Elementary School at Addison, New York just about on the Pennsylvania border. They have been following the goose odyssey for the past two years and have passed on their enthusiasm for the project to the whole school. They ask if it is possible that our route might pass over the school. I pull out the chart and find it is only a couple of miles off our planned route. I punch the coordinates into the Apollo GPS and assure the Abbeys that if the weather is right and we fly on time we will pass overhead. We are airborne out of Jalamtra at 7:40 A.M. and right on 8:00 A.M. we are over Addison Elementary School. Seven hundred kids are peering upward through the morning mist as our strange entourage of wings slips southward. Seeing all those schoolkids in the valley is as thrilling for Joe and me as I was told later it was for the kids. Our timing over the school is a little early. Kids pour out of their yellow buses wherever they stop, some not yet at the school, their drivers stopping on the street to let them witness our passage, while others already at school are out huddled in the tennis court in the cool morning air. How many centuries have people stopped their morning hustle to look up and rest a moment while on high a great skein of geese goes honking south? I could see the earliest North Americans emerging from their longhouses standing in the quiet of dawn, the passage of the geese a moment of reverie. Early settlers stopping a team of oxen, Civil War soldiers pausing while breaking camp.

On we fly. Roy Hughes has a wide, smooth, 2,000-foot airstrip situated on a hilltop near the Tioga reservoir just across the Pennsylvania border. Our original plan had been to fly direct from Jalamtra to Trout Run near Williamsport, a distance of about seventy-five miles. We worried a lot about Trout Run, deep in a valley in some of the most rugged mountains of the Appalachian chain. If there was any wind we might find ourselves in air currents that are unmanageable in our featherweight craft. The Hughes' strip is halfway to Trout Run, making it a perfect stop-off where we could wait out the midday winds and then cruise on to Trout Run in the calm of evening air.

The flight from Addison School to the Hughes' field is spectacular, the land smoothed by time, some of it covered in autumn golds and reds, green or light-brown meadows, flowing scarves of silken mist. Often we see whitetail deer standing silently watching us pass or bounding off into the cover of underbrush. Crossing over one hill we stir up a dozen or so wild turkeys along with a couple of whitetails. Goose speed certainly is the most beautiful way to absorb the autumn visions of northern Pennsylvania.

There is always a quiet period after we land. The birds, tasting the grass of the new airstrip, stick close to Joe and me as Roy and Betty Hughes bring tea and coffee. We all rest until the ground crew and curious neighbours arrive. When the pens are up and the geese secure, there are, as usual, reporters, then brunch in the local town diner. Once our identity is revealed as "the goose guys", there are always questions on what we are doing. They are always very welcoming people, the ones who gather, seeming to embody the essence of nature lovers, aviators, and the small-town attitude.

Later that day we are rolling down the Hughes' strip with the coordinates of Trout Run punched into our navigation systems. Directly ahead is one of the first high ridges of the Appalachians. There is no way we can get the birds straight over it. Veering east, we find a narrow valley leading south. With some desperation we work at getting the geese to form on the wing. Then we subtly increase the climb with the hope that by the time the valley runs out we will be high enough to clear the surrounding ridges.

It works. We worm our way up the most gradual of the valley estuaries until we break out on top to the great vista of those age-old mountains. Ridge and valley, ridge and valley, the mountains become more rugged and angular. Beneath us Route 15 weaves back and forth, following the deep valleys in a general direction toward Williamsport, while we pursue an almost straight line toward Trout Run. Thirty-six hundred feet over the mountains of Pennsylvania the colours are outrageous, the valleys full of mist and the lowering sun adding its share of drama. With thirty-eight birds in a perfect line off the wing, I am torn between pride of achievement and the burden of responsibility. I am awed by the spectacle and stunned by its beauty. I keep repeating to myself,

"Look where you are! Look what you're doing!"

At almost three thousand feet we round the last mountain and Jim Fink's Trout Run airfield is revealed nestled in the valley beneath us. By now Don Lounsbury is talking to us on the radio and reports light winds. We spiral down and down, orbiting the field within the valley almost at treetop. Joe is flying about three feet off the runway. With a bird right under his wing he's afraid to land because he doesn't know where the bird is going to go, so he flies around again. When he finally lands he has two birds right behind his wing with their heads above and their bodies below! He speeds up and they land behind him, the rest of the group following. After landing, the geese seem somewhat skittish, as if they want to take off again.

Jim Fink, the owner, comes out from under the dump truck he is fixing and welcomes us once again to "Finkhaven". I find a hose and get some water organized for the birds while Don and Joe attempt to keep them herded near the hangars. Still, we feel there is something wrong. The birds' body language shows that they are wary, necks stretched up. I spot a cat slinking off in the bushes, but it's too far away to scare the geese, I figure. What is troubling the geese? Just then the motor home and truck arrive. Geordie and Kirk soon have the pen in place. They have that down to a fine art. We start leading and herding the birds into the pen, when one just takes off and flies away to the south. Kirk nonchalantly states we have just lost a bird. The pen is closed and after four or five recounts there are still thirty-eight birds. Unknown to us we had picked up another wild hitchhiker, no doubt why the geese were skittish.

At the Trout Run Hotel and Bar, they know us from previous stopovers. The menu has not changed. I order the same as last year; we drink a couple of pitchers of beer, which consolidates our fatigue. So much seems to have happened in such a short time that we feel we have been travelling for weeks.

In the morning Joe gets airborne first to test the air. I lead the geese out and we head down the valley toward Williamsport, following a new, unopened four-lane highway. We break out into the Susquehanna Valley west of Williamsport and cross the river, a perfect mirror without the slightest ripple. The forest ridge ahead of us, called Bald Eagle Mountain, looks ominous. We swing east and again enter a southerly valley, finally getting the birds up and over the first ridge only to find a small valley and an even higher ridge, White Deer Ridge, facing us. It is rugged country with little sign of habitation. I swing west, trying to work the birds up diagonally. We pass over Miller's field, the only clearing in the valley, where we sought haven last year when head winds and turbulence caused us to put down. The birds have not climbed well enough. Joe takes the lead and swings them 180 degrees to the east and again climbs on a slow diagonal up the face of the ridge. It works, and after three or four miles we squeak through the ridge-top turbulence, the geese now down in the tops of the trees. We are just a little higher. The ridge now drops off sharply. The altimeter shows us about 1,500 feet over the valley floor, which has opened up into a great array of autumn fields and little farms that decorate the lesser ridges now well beneath us.

For a short time there is a sense of relief as we bask in

the luxury of this safe altitude, the geese now perfectly formed off the wing. An autumn haze removes us from the earth even more. In contrast, the sharp definition of this airborne ribbon of birds gives a surreal feeling, particularly as the birds are now bracketed fore and aft by buzzing white-winged machines eight times their size. But our reverie doesn't last. The birds gradually lose altitude and we arrive at another ridge we must work them over. Our ground speed is showing almost forty while airspeed is about thirty-five. For the first time we have a little help from the wind.

It is not long before we make contact with Ken Roadcap, who has an airstrip at Middleburg, a standby stopover. However, we are making miles now and opt to carry on to the Yosts' field in the little village of Sherman's Dale. There is now a high, thin overcast. The air is smooth, the flying easy. Our stop at Sherman's Dale is short, but wait, Igor is missing. Neither of us saw him drop out. We debate whether to proceed or stay. We are torn. The weather is perfect. We could fly right to Airlie, but we wonder. Perhaps we have overtired the geese. Or was it just a fluke and we lost a weak goose? It is a tough call. We opt for carrying on to Gettysburg, a distance of thirty-five miles. With the slight tailwind it will be less than an hour's flight and the ridges should be lower from here on.

Just before take-off a local TV reporter and cameraman shows up at the invitation of Mrs. Yost. The reporter is wearing a neat conservative suit, her hair like it was cast in place. She has not a clue of what we are doing. It takes time to get her up to speed.

That over, we get airborne and are now in the Blue

Mountains and I find I was wrong about the ridges betting easier. It becomes another struggle to climb over three more ridges, more rugged, pristine forested country, with the birds again just at treetop, scratching over the lowest points we can find in the ridge. Again we take up our routine of relaying leadership, one of maintaining a view from the rear, the other flying above and ahead of the geese.

Finally we are out of the ridges and pass over Highway 81. The contrast is extreme. At one moment we were practically scratching through wilderness treetops, now we are flying over suburbs and Wal-Marts. We pass over a large school complex. The soccer game beneath us stops as the players stare upward. The high ridges now have given way to lesser hills, and rows of fruit trees. Passing to the west of Gettysburg we put down at the home of Liz and John Brown, marvellous hosts. We are greeted by a pair of Amish gentlemen fascinated by the birds and the aircraft. Joe has worriedly counted the birds in the air. A recount on the ground confirms there are two more birds missing, Fleck Neck and a numbered goose. We have pushed them too hard. It is time to give them a day off. Even Joe and I are feeling the ache in our shoulders from four days of working the wind. There is little we can do about the missing birds other than call back to the previous two strips and see if they have returned. The whole crew is saddened at the loss. We hope they are all right, sitting on some lake in that high wilderness — or could they have landed at a crowded shopping centre? We never found out.

Sunday morning, October 16. There is ice on the wings, hard to remove. By 7:45 we have said goodbye to the Browns. Danny, their precocious five-year-old who had been demonstrating his ability to drive a tractor the day before, is saddened by our leaving. We get the birds to seven hundred feet and parallel the ridge that runs southwest, eventually to become the Blue Ridge Mountains. For a short time we fly over a corner of Maryland, passing to the west of Frederick over a densely populated countryside, a sunny Sunday morn with few people stirring.

Edging closer to the ridge, the birds have dropped down again. We cross through at Point of Rocks where the Potomac makes a gap. Now in the rolling horse country of Virginia we pass by opulent manor houses and over a large equestrian event. One lady in a light blue robe waves to us from the balcony of her pillared mansion. The homes here are in great contrast to some we have passed over in the Appalachians to the north.

Ten minutes out of Airlie the birds must sense they are near their destination and are now down to treetop. As we cross Highway 66, the air that has been smooth erupts into severe turbulence. We are kicked all over the sky, the worst I have ever experienced. We cross the home of Till Hazel, a mile or so north of Airlie. A few hundred wild geese rise from his pond. We are being bounced around badly. Our geese have lost their flowing line and are in a ragged bunch between Joe and me. It appears for a moment that they are going to dissolve into the wild flock that has risen to almost the same altitude. Joe and I call to them frantically.

They stick with us. Soon the airstrip at Airlie is dead ahead. A gathering awaits our landing. The air is still aflutter but we get the birds and aircraft down safely and are soon sipping tea and shaking hands while the geese wander among the welcomers, greeting them in their own goose way.

At some point there Joe and I look at each other and feel good. We had repeated unerringly the same flight as last year, proving to ourselves that we were not a one-hit wonder. Airlie is halfway to our new goal of South Carolina, where there are alligators but no goose-hunting season. We look forward to planning this next expedition while geese and crew take a rest.

CHAPTER FOURTEEN

The Last Leg

It was a relief to be at Airlie once again. We could all relax a little and enjoy fresh beds, clean clothes, and showers. Five of us living out of a twenty-year-old motor home supplemented by a tent gets a bit trying. We were missing the three geese that had dropped out while crossing the Appalachian ridges. Their absence concerned us. Geordie and Kirk had grown attached to the missing birds, particularly Fleck Neck and Igor, but we knew that they would probably join up with a wild flock and survive O.K.

Our general plan that year was to take the birds on to somewhere in South Carolina, but the timing had been left loose. Dr. Sladen and his team of biologists had an agenda they were promoting. They would like to keep us and the birds at Airlie for a month to do some testing on neckbands and fit tracking radios. Joe and I wanted to establish landing spots farther south as we had done the previous year through New York and Pennsylvania. A couple of weeks' layover, we felt, was enough for both us and the birds. The issue was further confused by a commitment Dr. Sladen had made for us to deliver a lecture to the Explorers Club in New York City on November 28. If we flew the geese on to South Carolina in mid-November we couldn't be sure of keeping the date in New York. Also, if we got the geese to South Carolina too early, the alligators would still be active and hungry for fresh geese. Also, the weather might still be too warm for good flying.

We finally agreed that we could continue our flight south after December 1. But first we'd have to lay out the route, so Joe and I, sans geese, headed out to do so. We figured that within a few days we could establish landing places and a good goose-wintering spot. It turned out to be not that easy. While there were many small grass airstrips, there was no comprehensive listing or plotting of them on available maps. Some were shown on aviation sectional maps, but locating them by road proved difficult. In two days we established only two possible landing spots. Not enough, so we chartered a Cessna out of Farmville airport in the middle of Virginia. Then in two days of flying by Cessna we had identified eight potential spots and had checked out three potential goose-wintering areas on the South Carolina coast.

Best of these was the Tom Yawkey Wildlife Center on South Island plantation, a twenty-thousand-acre wildlife preserve comprising three coastal islands, part of the Santee River delta about fifty miles north of Charleston, South Carolina. Bob Joyner, managing biologist at the center for eighteen years, greeted us enthusiastically. His tour of the island included a series of entertaining stories about the area from its days as a rice plantation before it became a preserve owned by the Yawkey family. His graphic descriptions of the ravages of hurricane Hugo led to a series of minilectures on the flora and fauna we were seeing. Quite satisfied that we had found the best over-wintering location and enough intermediate stops to suit our needs, Joe and I headed back north to make our date at the Explorer's Club in New York. This was pure culture shock compared to the mainly rural personalities we had met at our various stops en route with the geese.

The club, a mellow old establishment on East 70th Street (with Sir Edmund Hillary as honourary president), dates back to the likes of President Theodore Roosevelt and Admiral Richard Bird. Dr. Sladen is a gold medallist member of the club, having received the award for his research work with penguins on several Antarctic expeditions. In this haven of like-minded people, we were well received.

By December 1, we were back in Airlie ready to continue south, to find that in our absence another little drama had taken place. After landing at Airlie, we had thought it might be a good idea to call radio stations and newspapers in the vicinity of where we had lost the three geese in Pennsylvania to see if we could find them. Kirk took on the project. Within a few days we had a call from Gail and Bill Hunter from Lockhaven, Pennsylvania, a little west of the route we had flown. They had sighted number 38 goose, Igor, mixed up in a local flock that came to feed outside the Hunters' home. We made ready to head north by truck to bring Igor to Airlie. There was no need! The Hunters were so excited by the project that they coaxed Igor into a portable dog kennel, loaded him in the back of their Bronco, and drove him the two hun-

34 The airborne canoe we call Goose Leader.

35 The plywood gander we used as an imprinting tool.

36 A young goose exercises its wings prior to flight.

37 Into the dawn sky.

38

39

40

41

38 One of the flock catches a free ride on the Riser's leading edge.

39 Approaching Lake Ontario at 700 feet.

40 The chase boat stayed beneath us most of the way.

41 We cross the Susquahanna west of Williamsport.

42 From above, the gods-eye view of what we do.

43 On the wing over plowed fields.

44 The welcoming committee at the Tom Yawkey Center in South Carolina.

45 Dr. Sladen imparting goose wisdom.

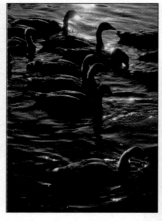

46

47

46 Pre-fledging goslings frol-
  icking in the pond.

47 Geese sense danger,
  while I remain oblivious
  to the alligator lurking
  beneath.

48 North Carolina sunrise.

48

49

50

49 North Carolina sunset.

50 Thurman Batchelor inspects his overnight guests at Lake City, South Carolina.

51 Bill Lishman and Bill Sladen, happy to be on the ground at last.

51

Greg Hewson

53

54

52 Goose landing
— reverse thrust.

53 Side by side, Joe
Duff and I lead
them south.

54 Just before we
land at the lotus
shrine of Swami
Satchidinanda.

55

56

55 Morning mists
transform the terrain
of North Carolina as
the geese fly south.

56 Another one of
Joe Duff's shots in
the golden light of
dawn.

57

57 The day-care kids
come to visit the
geese at the home
of Dawn Wallace in
North Carolina.

58

58   Geese off the wing at sunset.

59   The crew of Operation Migration 1994: (left to right) Joe Duff, Geordie Lishman, Kirk Goolsby, Chris Fox, Bill Lishman and David Woodhouse.

59

dred miles to join us at Airlie. Igor, according to Kirk and Geordie's observations, until then had been our head honcho goose. We were sure he would be welcomed back by the rest in grand style.

The Hunters arrived with Igor about noon. We all gathered at Raven's Hollow, the part of Airlie where the geese hung out, to watch Igor being greeted royally. We carried him out to the flock in the dog kennel and opened the door. He stepped out hesitantly — and a total reversal of what we expected took place. He was driven off by member after member of the flock. Igor was an outcast. From leader of the flock, he'd landed at the bottom of the pecking order, a pariah. We did have one clue. The rest of the flock had been neckbanded in addition to their leg bands that they had worn all summer a couple of days prior to Igor's return. He had not been neckbanded yet. Was it because the rest were banded and he was not? Had the goose that had taken the lead in his absence organized an anti-Igor campaign? It's hard to know what goes on in the minds of geese. Within a few days, after Igor had been neckbanded like the rest, he was assimilated back into the flock but not as leader, his much anticipated welcome-home party a total bust.

On December 1, then, with our global positioning system pointing us southwestward after the six-week layover, we were once more airborne out of Airlie, this time with a thirty-six bird entourage. All the leaves had gone, the brown grass of the Virginia fields pastelled white with the morning frost. The geese, grown lazy with their rest at Airlie, had no inclination to climb. We dodged to the west of Warrenton to avoid overflying it at 100 feet and picked our way carefully over open rolling countryside.

Sometimes in the first few miles we had to go around trees rather than over them. Gradually we got the birds up to about 150 feet. Deviating around the town of Culpepper, we were a little off course for our first planned stop at Bob Snow's airstrip Snow Hill, slightly southwest of Charlottesville, Virginia. With about ten miles to go and after barely an hour in flight, Joe and I had traded leadership several times with the air full of wild flocks of Canada geese. Joe was leading the flock and we were skirting to the east side of a very prosperous-looking white-fenced horse ranch. One goose had been acting up, breaking out of formation and flying low. It was obvious the birds were tiring. Directly ahead was a large pond surrounded by trees on three sides. Off to the west and ahead of our entourage on our long approach to the pond was a flock of fifty or more wild geese. They passed directly in front of Joe and the Ultrageese. I was behind and above. Our geese broke and started following the wild flock. I peeled off in pursuit. Joe carved around steeply to the right. The wild geese landed in the pond. Our birds overflew and started a circle. They definitely had a landing in mind. I caught them over the trees to the north and flew right into the middle of the flock, trying to get them to form on my craft. My airspeed was a few miles an hour too fast and I overshot the flock. Glancing over my shoulder, I saw Joe swooping in for another intersect. Again I attempted to pick them up. All the time, Joe and I were keeping our transmit buttons hot with quickly planned strategies. Nothing worked. The geese circled lower and lower. There was one horse in a paddock close by. It seemed unconcerned. Other horses were pastured a little farther to the west, seeming unperturbed. That was not the case with a woman who emerged from an office next to the stable, frantically waving us off. There was nothing we could do. Our birds were on the water, mixing it up with the wild flock. It was the first time they had landed at a strange spot without our leadership.

Climbing over the hill to the south, I spied a vacant pasture with a trail running diagonally. It was on top of the hill and a couple of hundred yards from the pond. While Joe attempted one more low-level run over the water to get the geese airborne, I landed on the farm track and taxied up to a small building on top of the hill. A couple of minutes later Joe joined me on the ground, his low-level run to no avail. We sat there for a moment in shocked silence, neither of us with any idea of what our next move might be. We soon found out. Within a few minutes a small pickup arrived and a young guy in a baseball hat informed us that we were trespassing. I attempted to explain what we were doing on the field, but did not sound convincing even to myself. I could see in his eyes that he felt the same.

Flying with geese? What the…? I took another approach and asked if he would take us to meet the owner. Fine. We jumped in the pickup and headed back down our commandeered landing strip. The ranch was beautifully kept, white-painted oak fencing dividing each pasture. A large white manor house surrounded by ancient oaks sat on a well-landscaped knoll.

We bypassed the house and were taken to a small, quaint office building set off from the equally well-appointed horse barn not far from where the Ultrageese were socializing with their newfound friends. Charlie, the pickup's driver, directed us into the office and announced that a couple of small planes had landed in the back pasture. The woman behind the desk who had been out waving us off was obviously distraught. She did not yet see Joe and me. She said she had read our aircraft numbers and put a call in to the FAA. I had worked my way in to the office by then. "Well you don't need the number," I said, "You now have the pilots."

Joe apologized profusely for the intrusion and I made an attempt to explain the circumstances of our landing. She took some time to calm down and pointed out that she had extremely valuable horses that must not be spooked. We explained that we did not overfly the horses, only the one near the pond which we had observed taking in the scene with a large degree of serenity. Eventually this woman, Miss Peggy Augustus of the old Keswick Horse Ranch, remembered that she had seen a TV program on our goose flights. She then warmed to us and the project. Soon we were on her phone to our ground guys, who were also invading the ranch.

But the crisis ended well. The pen was set up and we were welcomed to stay the night. The Ultrageese II socialized for a short time with the wild flock that had lured them into the Old Keswick pond of Miss Peggy, but they soon answered the call from Geordie and Kirk and waddled out to stay overnight in their familiar pen. The next morning, a few of Miss Augustus's friends were invited over to witness a routine release and take-off as we joined the flock in mid-air. Once again we took up a southwesterly heading as if it had been a scheduled stop.

Each day held new drama either in the air or on the ground. Our next planned landing was Yogaville, situated on the south bank of the James River in hilly country.

It is the ashram of yoga master Swami Satchidinanda. We had a slight head wind and the air was bumpy. Five miles out, we made radio contact with our ground people who had arrived ahead of us. As usual the birds were treetopping it and we were having to stay above them to avoid being trashed into the trees. For a short period, we followed the meandering of the James, then crossed over the last hill to the wondrous view of the swami's lotus shrine and the airstrip that paralleled the river.

The shrine lies in a Shangri-la valley. The image of it nestled there in the shape of a lotus flower was striking in contrast to the Virginia architecture we had been passing. The geese were on my wing. I flew a wide circle, keeping the lotus shrine at the centre. Almost over the shrine, I was hit by a rude piece of turbulence that had rolled over the southerly ridge. It sent me sideways. The geese broke out of their formal line. Struggling the wing back into shape, I set up an approach to land. Joe crackled through my earphones politely requesting that he might land first because his engine had just quit. No debate on that. Soon we were both standing amid the geese on the roughly mown runway, obviously now used as a hayfield rather than an aviator's mecca. Don, leading our ground crew, had just beaten us to the airstrip and was now talking the rest of the ground crew into the strip by radio. We paused, thinking about Joe's engine and how polite it was to quit at the correct moment.

The lotus shrine is something to experience. Dedicated to truth, peace, light and the acceptance of all faiths, it is a meditation centre — all welcomed. The shrine itself is an amazing piece of architecture formed in the shape of a lotus flower. A landscaped formal processional leading from the entrance arch dramatizes its setting. It had the flavour of a mini-Taj Mahal. After the daily meditation at noon, we were invited to lunch at the ashram. Wholesome vegetarian food was consumed in meditative silence by all. All except Joe. He had headed out by road to Scottsville for a burger, and more important, new spark plugs for his wayward engine.

I had met the swami twenty-two years before, during our struggle to stop the building of Toronto's second airport. He had travelled to Ontario to open our anti-airport environmental festival, Earth Days. Today, however, he was nowhere to be seen, at least by us. The next day was his eightieth birthday. A great celebration had been planned in Charlottesville, thirty miles to the north and the swami was busy with the preparations.

After lunch we rested, planning to carry on in the late afternoon. Then I discovered that the ashram closed at 5:00 P.M., gates shut and locked. It would be closed all the next day because everyone from the ashram was going to the birthday celebration in Charlottesville. We had a few moments of panic, wanting to get airborne and on to the next stop. We rushed to get everything ready for launch just before sunset but to no avail. A glitch in my radio rendered our communications useless and we were losing the light. Fortunately Shiva, the member of the ashram in charge of maintenance, came to our rescue and arranged for us to stay overnight.

At dawn the next day we found frost on the wings for the third day in a row. That was a real pain — we'd thought we were by now far enough south to avoid such frost. Clearing the wings was like scraping the ice off a couple of dozen car windshields that you can't quite reach. Joe became totally frustrated; I won't repeat his comments in print. The clean-up delayed us a full half-hour of perfect flying, but eventually we struggled the geese into enough altitude to leave the James River Valley with its giant lotus flower behind and set our sights on Big River Ranch near South Boston, planned as our last stop in Virginia.

The terrain levelled out somewhat but the air got bumpier and bumpier — uncomfortably so. After fifteen miles we put down at the home of Tim Hoag for a brief cup of tea. After a half-hour layover, I got airborne and found that the air had smoothed out a little. The Hoags helped us get the birds aloft once again, but that took three attempts. Then, with a slight head wind, our ground speed was only twenty-four miles an hour. We plodded on, low and slow over mile after mile of pine and oak forest with no place to land save the rare clearing. We began to worry about making Big River, so changed our course for the nearer alternate airstrip of Jim and Slovi Smiley near the village of Clover. We were able to make garbled radio connection with the ground crew and redirect them to meet us at the Smileys' strip. Somehow our GPS coordinates were slightly off. We had the darnedest time finding the place in mile after mile of dense woods, with no place to put down. We began to worry about fuel and the condition of the birds. For the first time, I detected a note of panic in Joe's voice. We got Kirk on the radio. Geordie and he were in the truck, having difficulty finding their way in to the airstrip on the ground.

There was a sharp dissertation between Joe and Kirk: Joe: "Can you hear us?" meaning could Kirk hear our engines and say where we were from the field?

Kirk: "Yeah, I can hear you fine," meaning at least that the radio transmission was working well.

For a moment there was plenty of tension while we orbited, desperately attempting to find that necessary piece of grass hidden among all those tall pines at Clover.

After several half-mile-wide circles, Kirk had us in sight and directed us over the field. A couple of circuits and we were all down safely, greeted by Slovi Smiley.

The Smileys have a lovely spot. Jim, a captain with Delta Air Lines, showed us his three aircraft, all in immaculate shape. Our desire was to push on to the Comptons' Big River Ranch, where Jim and Sandy Compton told us by phone that they had a guest house and hangar available for us. The day had turned overcast, with a threat of rain called for the next two days. At four-thirty, with only an hour of daylight left, we climbed out of the Smileys' field, our geese attached, and within half an hour we were setting down at Big River Ranch. On the banks of the river the setting was stunningly beautiful even in grey, overcast twilight. Jim and Sandy made us welcome. What a spot: a hangar that both the aircraft could fit in and, attached to it, a fully furnished apartment at our disposal. The geese had it good too, with a river to frolic in. The weather did close in and rain began. It could not have been a better spot for that to happen.

Jim Compton had been a crop duster for over thirty years, laying waste to that old adage that there are no old, bold pilots. Jim regaled us with story after story of low-level exploits punctuated with descriptive phrases like "he could fly that plane like a monkey handles a peanut" or "his aircraft was so well polished it was like mercury on a dime". Sandy had been Jim's secretary for five years until Jim realized that the best thing was to marry the lady. That was thirty years earlier. Sandy was very enthusiastic about the geese. Her hobby was to watch the bird life that frequented their area, mostly done with binoculars. This was her first experience with geese so close at hand.

We were kept at Big River two days by bad weather. We really had to rough it, says I with tongue in cheek. On the Sunday the Comptons had us accompany them to Ernie's Restaurant in South Boston for a great meal, followed by a flight with Jim in his retractable-gear Cessna. This was not just your ordinary scenic cruise. It was a sample of Jim Compton handling that craft like a monkey handles a peanut. At below treetop, we streaked down Big River at 130 miles an hour banking through the curves at boat level! Quite an extreme speed for us who were used to plodding through the air at just over jogging speed.

The weather improved. We could fly again. Big River Ranch was still not halfway between Airlie and our planned destination near Charleston.

We carried on and landed at the home of Mike and Debbie Miles. The same day, we flew another fourteen miles to put down at Eagle's Landing, a strip hacked out of the bush by an American Eagle pilot. I had left a message on his phone that there was a possibility of us landing there, but on that bush strip there was no one home but us and the geese.

A little earlier, Joe had been having trouble maintaining straight flight. Pilot Dave Woodhouse tried it (second opinion). On climb-out the engine sputtered and coughed a bit. David quickly brought it back down. We huddled around trying to find out what was wrong. There were so many redundant systems on the craft, it seemed inconceivable that there was a problem. We thought it had to be fuel starvation of some sort. Joe pulled out the tiny filter that is the last-ditch fuel filter, situated right on the carburetor. It was clean. Then we discovered a blockage in the fitting that held the filter. A bit of melted fibreglass strands stopped by the filter had prevented the engine from getting fuel flow at the full-throttle situation, but allowed enough flow at cruise settings. We also found a misplaced rib in the wing that had been incorrectly installed. This was easily missed on our preflight walk-around, but it did make the wing pull slightly to the right. With both problems corrected, Joe test-flew a full-throttle climb-out and a hands-off cruise. He was happy once again. But I began to worry about fibreglass fibres in the carburetor on my aircraft. It was a different type of carb and did not have the same filter. I checked the in-line filter. It seemed fine, ready to go.

The next day's flight was truly an extreme experience. We had no ice on the wings and were ready for the air before it was light. There was a little ground mist, so I took off solo for a quick reconnaissance flight. The visibility was a couple of miles or better with patches of ground fog, nothing that we had not been able to handle previously. Joe joined me in the air. We did our usual air pickup of the geese and set our heading for the Wallace family airstrip near Aberdeen, North Carolina, on a spectacular morning. The sun rose over the patches of ground fog, picking up a variety of colours. But as we flew on, the spotty bits of fog turned to a soft, undulating blanket of white. Looking straight down, the forest and fields were hazily discernible through the waving mist. But all around us was a most wonderful world of white with islands of trees protruding here and there, with the sun low in a pumpkin-coloured sky.

It was the most spectacular flight of the year. There was so much moisture in the air that even my glasses kept fogging over. The geese in that golden pastel light of dawn stood out in breathtaking contrast, a vision of primal beauty. The major frustration was that we could not capture the image either on film or video — both cameras were too fogged up to work.

While we were caught up in this sea of wonder, flying from island of trees to island of trees, each clump standing there skeletally outlined like some great pop-up out of an endless page, now and then we could catch glimpses of a farm or highway through the mist. Anyone down there who saw us would hesitantly wave and we would wave back as we slipped by, a ghostly apparition of strange craft trailed by a couple of hundred feet of honking geese.

I put myself in the place of the few who saw us from the ground, imagined us appearing out of the fog only a couple of hundred feet above and then disappearing within seconds. I would love to have heard them describing this phantom vision that startled their day.

But the fog was not getting thinner. We were sure that it would burn off when the sun got a little higher. Joe and I discussed the eventuality of landing at Wallace if we still could not see the ground. We could definitely find the field with the GPS, fog or no fog, as long as we had a horizon to keep us in level flight. When we had travelled through there in the Cessna a month previously, I had punched in the exact coordinates on the GPS that would get us directly over the field. So that part was O.K. or almost O.K. Also, if we turned back there was no assurance that Eagle's Landing would not be fogged in too, with the ground crew miles away.

It became a real worry, up there in all the whiteness. Taking stock, we had four hours of fuel. I assured myself that in the worst-case scenario, the ground crew would be able to find the Wallaces. We could orbit above the field, the geese could be called down by Kirk and Geordie, and when the fog burned off we could land.

We were cruising at a ground speed of twenty-eight miles an hour and there was only fifteen miles, or just over half an hour, to go.

Then out of the mist, in a patch of almost clear air, appeared a brand-new freshly paved runway. The rest of its surroundings were still under construction. It slipped under us so quickly, and we were so enraptured with the day, that neither of us was bright enough to take a moment to mark its location on our GPS.

We flew on, with the image of that airstrip as a safe haven sitting worryingly on our memory. Both of us were biting our lips, thinking silently that we should have taken the opportunity to land. Hell, it was only half an hour and a bit to the Wallaces....

Ten minutes later the fog was rising around us, the earth barely discernible. A layer of cloud thickened above us and the horizon ahead began to blur beyond recognition. To the east and west, it looked no better. Our world was disappearing. There was strain in our voices; there was no debate. We executed a slow 180-degree turn, geese still faithfully in tow....

Quickly now, what was our heading? 208 degrees. What was the reverse heading? 208 minus 180 — 028 degrees. My dyslexic brain under pressure struggled with the simple arithmetic. It was three or four miles back to that paved airstrip. Would we be able to find it again in this endless soup? If our heading was off a little, we could fly right past it.

Zero twenty-eight degrees. But there was a wind from the southwest. Let's try 027 degrees. Fingers crossed, we buzzed on in radio silence.

Our ground speed with the tailwind was almost forty. The strip should appear in maybe five or six minutes. Would the blanket have moved in and covered it too?

Silence, eyes straining into the undulating whiteness ahead for that open spot of haven. Five minutes. So long... like a child waiting for Christmas. Then relief! There it was, still clear! A pristine piece of pavement ninety degrees to our path, kept open by some guardian angel knowing we would return. The strip was brand new, not a mark on it, with one lonely airplane tied down at its edge and a few bulldozers and other bits of earthmoving equipment scattered about.

I did a circuit with the geese still off my wing and landed, hoping the birds might follow straight in. Joe stayed up and orbited in our normal landing routine. I rolled off the runway onto hard-packed, rust-coloured mud.

Untangling myself from the aircraft, I call out to the geese. Joe lands before the birds and we watch with great glee at how gingerly they make their first pavement landing. Some think it is water and try a feet-down water-ski landing and stumble a bit while others, realizing they are landing on hard stuff, fare better. Still others land sloppily in the mud.

Soon people who had seen us land arrived. We were in Carthage, North Carolina.

That evening, under a clear and bright sky, we landed at the Wallaces'. Dawn Wallace runs a day-care centre. We were greeted by about ten five-year-olds. There was an immediate and mutual fascination between our flock of geese and these little people, a truly entertaining interaction between the two groups of juveniles. Dawn's husband, Bunny, had the barbecue going and had brought in half a bushel of fresh oysters from the North Carolina coast. What a feast! A great celebration, and not only of the oysters.

The winds were too high the next morning, but by mid-afternoon we were southward bound once again. Crossing into South Carolina we landed at the crop-duster strip of Sherman Hanke south of the village of Clio (pronounced Cly-oh). He is called Hank for short for, as he says, the name Sherman is not overly popular in the South. He adds to the repertoire of crop-duster knowledge that we learned earlier in the week from Jim Compton, and shows us around his hangar, which holds four huge biplanes, two of them ancient Steermans still reeking of pesticide. He has been flying them since 1946, and tells us that it does not pay to fly anything lighter for you need big iron to cut through the power lines that you may accidentally hit on occasion. In his back hangar is the Stamp biplane that was once the star of the Hollywood movie *High Road to China*.

We now had just under ninety miles to finish our mission. Dawn brought a full overcast sky but nothing that would stop us heading on our way. In the air we could almost taste the ending to our odyssey. We could almost do it in one run except that the terrain beneath us was most uninviting — not marsh but mile after mile of swampy bush. If there was a middle of nowhere this was it.

I wondered again about those little fibres of glass that we took out of Joe's carb and how they might affect my carb.

We put down next at the home of Thurman Batchelor just east of Lake City, South Carolina, meaning just to stretch our legs and top up the fuel tanks before going on. But while we were still on the ground the odd bit of rain started to spit down. I did a short reconnaissance flight to the south while Joe sat with the birds. On return, my news was that probably we were on the ground for the rest of the day. I called ahead to resident biologist Bob Joyner at our destination, the Tom Yawkey Center. He was expecting us within the hour and had a number of people who had made the trek out to the island as a welcoming committee. I told him there was little likelihood that we could make it. An hour later the sky seemed to clear and looked O.K. for flying. We suited up and I called Bob again and told him we could make it after all. No sooner had I hung up when I got a call from Don Lounsbury, twenty miles down the road, reporting heavy rain squalls. We cancelled again.

Saturday was our tenth day out of Airlie. We could have walked and covered the same distance. We awoke to solid fog that morning, finding that about twenty people had turned out in the murky dawn to witness our departure, which the fog delayed. T. C. Batchelor, Thurman's brother, drove the whole crew into Lake City for breakfast. On our return the ceiling had lifted a couple of hundred feet and Thurman was out taking people for rides in his Cessna 150. Checking the weather ahead it was a bit iffy, but Georgetown, just north of the Yawkey Center, was reporting clear and Bob Joyner was reporting clear at the Center. We were airborne at 11:00 A.M. Scratching along under low overcast that was sometimes down to a hundred feet, we passed the odd radio tower that disappeared up into the grey. Twenty miles out, the horizon became a bright streak. It got lighter and lighter until just north of Georgetown we broke out into clear sunlight. Georgetown lay directly in our path. The image of heavy industry dominated the skyline. A paper plant and a steel mill spewed great palls of vapours into the sky directly in our path. We wormed our way between these man-made clouds, passing over a ship tied up in the harbour, the picture in sharp contrast to some of the great expanses of forest we had crossed in the days past.

The geese were well locked onto Joe's wings. I was behind and a few hundred feet higher, attempting to get an overview of our destination about eight miles ahead. We flew to the east of Georgetown airport, informing them by radio of our passage. There was a large fog bank ahead. From Joe's altitude he could not see beyond it. I put on the power to climb above the cloud. At full throttle the engine sputtered and wouldn't go to full RPM. I backed off and the engine smoothed out. I was sure the problem was those little fibres in my gas. I quickly checked the GPS. It was three miles back to the

Georgetown airstrip, four miles to our destination.

Joe said, "We're turning back."

I was at a thousand feet, the engine running fine at part throttle but if I tried full throttle it sputtered.

I was satisfied that it was not going to quit, and had a road picked out within gliding distance off to the west just in case.

I convinced Joe that we should press on. It was the fog bank that was worrying me. I raised Don in the ground crew by radio. He was on the ferry going toward the island. A few miles short of our planned landing spot he reported fog. Joe could see only the fog. From my higher vantage I could see that the south end of the island was clear. I directed Joe on a heading around the fog bank and set up for a straight-in approach. The field was just one clearing in a forest of sixty-foot-high pines. There wouldn't be any second chances on this landing, with no power to climb back out.

I touched down and the birds did their traditional three go-arounds and came in with Joe to a greeting from Bob Joyner and Dr. Sladen who had driven down from Airlie for the occasion.

The flying odyssey was over. A great relief welled up in all of us. There was a large welcoming committee. Bob's wife, Cornelia, had a huge plate of homegrown shrimp and beverages ready for us. The geese walked among the crowd and were panting as if they needed water. We walked them over to a nearby freshwater pond, but strangely they would not go near no matter how hard we splashed and coaxed them. Finally I waded out into the pond to show them it was quite safe. "No deal," they seemed to say and stood well back. Soaked to the thighs,

I gave up and returned to shore. Later in the day Kirk spotted a sizeable alligator in the pond. What instinct from eons past I wondered had kept those birds from swimming into danger? Obviously a self-preservation instinct that had been lost in this human. Later that day I disassembled the carburetor on my aircraft and found a little brass mesh filter matted with a pad of minifibres. "Two close ones in one day," I thought.

South Island Plantation's history and texture is all that is Southern. While sitting quietly with the geese, you can almost hear the soft Southern voice of the plantation owners who used to summer on the island to escape the inland humidity and mosquitoes, or smell the rice fields from days gone by. When Tom Yawkey became the owner of these 20,000 acres of marsh, ocean beach, maritime forest, and pine land, he hoped he would "be able to do some good with it". He was a man of his word. Baseball and wildlife were his two passions. The Tom Yawkey Wildlife Center is a symbol of his dedication to protect wildlife throughout lifetimes of generations to come. To the wild occupants, save the chance of being devoured by a 'gator or a bobcat, it is a safe winter paradise in the midst of a number of old plantations kept as exclusive hunting preserves. The old rice fields have turned to marshes which provide refuge for countless waterfowl. Threatened and endangered species have found sanctuary in the forests and dunes. Miles and miles of undisturbed beaches and pastures of marshy grass give to me, and others, an intense feeling of nature that to some could only be described as peaceful while to others would possibly bring a feeling of loneliness.

It was December 13 by the time we were packed up to drive home, leaving our wards in the comfort of the Yawkey Center. We had wanted to fly the birds locally for several days but high winds precluded that option, except for one short turbulent flight where I was able to get airborne and lead them over to the goose pasture, an area of the centre that lived up to its name and would be the best habitat. Kirk set up camp to monitor the birds over the winter as the rest of us headed north to rejoin our families and friends for the Christmas season. It seemed that all things were under control and in order.

It was not so! Two days after our departure the whole flock went missing. There was little concern at first for we thought they would show up again within a few days. A week passed, no geese! Christmas came and went; no geese! Kirk was beside himself and searched all the local ponds. The newspapers and radio stations of South Carolina did numerous stories on the missing birds. Many people watched out for them and called in if they thought they had seen them.

Kirk found an obliging pilot and did numerous air searches in a Cessna. It was all in vain. There were many reports of flocks of geese but none of them wore the distinctive neckbands of the Ultrageese II. Then on the last day of the year, Crash, number 44, was located north of Myrtle Beach and was returned by truck to the Yawkey Center. He took up cohabitation with Kirk, who continued searching throughout January and February. All the efforts were fruitless.

There were all kinds of conjecture as to the demise of the Ultrageese II, for with no fear of people they could have easily met with foul (or fowl?) play. Dr. Sladen was sure they were in someone's freezer. A rumour circulated that they had been Christmas dinner for a bunch of bikers. By March none of us held any hope of ever seeing them again, for if they were alive someone surely would have seen them. It always hurt when we were asked about them. Mostly they had dropped from our conversation. Our sights were set on new goals to fly with sandhill cranes.

In early April, Joe and I sadly returned to the Tom Yawkey Center, picked up our airplanes, and trailered them back to Canada. On the long drive north, neither of us spoke of the geese at any length, but now and then we would reminisce about the flight south as we came on various features that we had passed over the previous fall and now recognized from the ground. We would catch ourselves searching the sky or looking carefully at ponds as we passed.

We were not home a day when Gavin Shire called from the Airlie Center in Virginia. Igor had been sighted by a New York State wildlife officer in a park near Niagara Falls. We sat in disbelief, Niagara was a hundred miles west of the route we had taken south in the fall and some eight hundred miles from the Yawkey Center, but it was true, the neckband number K738 belonged to no other goose. We jumped up and down with excitement. We could put a radio collar on him and perhaps if the rest of the geese were nearby, he would lead us to them. We made plans to go the next day to Niagara Falls and meet Kirk, who would fly up from Airlie with the radio collar.

The next day as Geordie and I prepared to leave, we were stopped at the door by the phone. It was a worker at the sod farm. He reported that there were twenty-eight geese wearing neckbands on the front lawn. Again for a second we stood in disbelief. In a flash Joe and I scrambled for the car and within minutes were jouncing up the winter-heaved driveway of the sod farm.

There they were, twenty-eight, count them, twenty-eight honking, gabbling, beautiful geese and one wild one. Where had they been hiding? What route had they taken north? We flopped down on the grass amid them and just revelled in the reunion, feeding them corn as the elation welled within. There they were, Egghead, Ringneck, Spanky, Spot, Peckerhead, Peppy, Coffee, Sam, O.J., Ogar, Eyes, Roman Nose, Homer, Bugler, Clunkhead and a host more. Twenty-eight out of the thirty-three that had made it to the Yawkey Center eight hundred miles to the south. Later in the day four more geese showed up, three wild ones and Fleck Neck, who had dropped out on our leg from Sherman's Dale to Gettysburg and had not been seen since that day in mid-October. Thirty geese now accounted for, but no Igor; he must be lost in Niagara Falls. O.K., we decide we will head out for Niagara and find Igor. Geordie and I arise the next day ready to brave the Toronto rush-hour freeways headed for Niagara in search of Igor, but again the phone rings. Four more neckbanded geese have arrived. Again we dash over to the sod farm for another reunion, and there on the lawn of the old brick farmhouse standing a little aloof from the others is K738 — Igor — nonchalantly picking up kernels of corn out of the newly sprouting grass just as if he had never been south. I guess he had always wanted to see the falls.

In Conclusion

"Joe," I ask, "why are we doing this? I mean, really, what's the point, why are we risking our necks? Why are we dedicating a major chunk of our lives to this wild scheme? Just for fun? For science, glory, adventure? To save the birds? Are we trying to help save the planet?"

Of course I understand why I wanted to fly with the birds. Since man saw his first bird, we have envied their three-dimensional freedom. Consider how long birds, almost unchanged in form, have flitted and soared overhead, taunting us with their apparent freedom! The birds were there in the sky and below it throughout the millions of years it took us to evolve from crawling beasts to erect, walking *homo sapiens*. A desire to experience the world as birds do — climbing, gliding, and diving through the ocean of surrounding air — has been in our hearts for countless millennia. It was not until this century that our dream to fly was realized, and that is barely a flick of the second hand in the clock of evolutionary change. It was only in the past few decades that we managed to fly at bird speed, and still we cannot even begin to approach the agility and freedom that birds enjoy in the air. So Joe and I are acting out age-old desires of flight that have their roots deep in the shared consciousness of our species, a spooky and mysterious thought that sometimes makes me take a step back in awe.

Exploring our ability to re-establish the migration routes of endangered bird species is probably, at least in part, driven by the desire to get more of that flying elixir. Risking life and career to experience the wonders of bird flight is a highly addictive pursuit, as is the quest for adventure in this age of crowded cities and number-crunching productivity. But there is something else driving us, the guilt many of us feel for what we humans seem to have done to our planet.

Astronomers say that your thumb held at arm's length against the sky covers fifty thousand galaxies. Each of these galaxies contains billions of stars, most of them as big or bigger than our own sun. Presumably, many of these stars have planetary systems. There must be billions of life-supporting planets out there. If, suitably humbled, we then turn our attention back to the daily news, we cannot help but suspect that this is a particularly screwed-up planet. And so much of the trouble — wars, pollution, environmental terrors — seems to be caused by one species, the species which prides itself on being the most intelligent life-form.

As individuals I have no doubt that we are the most intelligent beings on this planet. All living things strive to influence their environment to satisfy their needs, but all other species are directed by relatively limited and unvarying sets of instinctive behavioral patterns. Our ability to choose and to change has given us remarkable survival powers. But it has also increased the risk that we will trigger changes we don't fully understand. We are still missing the big picture, and while we are certainly intelligent, we are not yet globally wise. The human race is not yet house-broken, or should we say planet-broken? We are becoming aware that in our race for comfort, security, and affluence we have fouled our own nest. With that awareness comes fear and a growing sense of collective guilt.

Our sense of guilt makes us want to change things, to correct the wrongs. But there are many different visions of how we should proceed, many knee-jerk reactions. Thought and speech are two of humanity's most evolved talents, but as children of the Tower of Babel we have a tendency to squabble when clear thought and cooperative action are most needed. It is worrisome to see confrontation escalating every day — between loggers and "tree huggers", hunters and anti-hunters, developers and preservationists, "green" revolutionists and prophets of doom.

One of the least useful reactions to our problems is expressed by some so-called animal-rights or deep-ecology movements — that is, the simplistic belief that we can solve everything by "withdrawing" from nature. In fact, the impact of human civilization has already made itself felt in virtually every planetary nook and cranny. It is doubtful whether there is really any pristine wilderness left to go back to, even if anyone were seriously interested in taking up such an option. Like it or not, we are part of nature and we must face up to our responsibilities. We have taken our ship out into the open waters, and we cannot abandon the helm now. Sometimes this means we will act as predators, taking our share of what nature provides. Sometimes we will act as guides or protectors, providing safe haven and habitat for endangered species. But act we must, and our actions will have an impact on nature for better or worse. To pretend otherwise is nonsense, and it is dangerous nonsense.

I found myself caught up in this debate over the past two decades when Paula's creativity turned to knitting and selling beautiful garments made with fur. The central activities of our family have seemingly incorporated what many people consider to be two poles of a supposedly irreconcilable debate. I am often surprised by people who assume that I would never eat a goose; I prefer it to chicken. So there will be a goose on our table this Christmas, and NO, it will not be number 47. It will be a farmed variety this year, but still very much a goose. As I've said, I was raised on a farm. As children we made friends with the cows, but in the autumn they were slaughtered and went into the freezer. Sometimes we are predators.

Sometime ago I was called for my opinion on what should be done about the growing flocks of geese which are fouling many of Toronto's waterfront parks.

"Introduce foxes and coyotes into the parks," I suggested. (Shocked silence at the other end of the line.) "Or serve them for dinner." (Click.)

I don't recount this anecdote frivolously. I think it is important to get my cards on the table, in full view. My goal is not for the geese to become cult icons, like Brigitte Bardot's seal pups. My hope is that our work may help to restore some wonderful species of birds to their traditional habitat. Joe and I, with Dr. Sladen and his team of biologists, are in our small way adding to the scientific knowledge which may increase our understanding of how migration works. If this provides a better view of the larger picture, it may help us to understand our place in the pattern. It may teach us how to live in better harmony with our environment, to ensure our survival and welfare as a species — and as individuals. So this, perhaps, is what really propels us to fly with the geese. When I reflect on how we got to this point, I think of all the people in history on whose contribution we are riding, from Otto Lilienthal and Konrad Lorenz to Bill Carrick and Bill Sladen. I think of my family and all who have helped along the way. Joe and his friends who have endured the downs and shared the highs. I think of the insight gained into a little of what the bird feels.

On the other hand, I think about general aviation aircraft and I cannot help but note that for the most part the standard light plane is a crude, inefficient tool, flying boards driven forward by crude fans with minimal sensitivity to the air, whereas a bird, with its infinite flexibility of wing, not only senses the minutest air movement but compensates automatically. Standard light aircraft, to me, have stagnated in design. Compared to automotive design, the current light plane has advanced little beyond the model-T stage. From our bird-flight experience I see great scope for the development of wings that incorporate multidimensional flexibility, propulsion systems that are more efficient and less obtrusive, and guidance systems that incorporate automatic avoidance and simplified all-weather capability.

Not least important, this whole project is (so far at least!) a good-news story. It is providing intimate contact with some exquisite natural systems and a ray of hope in the litany of depressing political, economic, and environmental catastrophes. I believe that this is why so many people have supported us and taken such interest in our quest.

Many seasons have passed since that September morning when, in a brief moment, I first flew wing-to-wing with a massive flight of ducks. Along with the seasons have come many changes in direction. Joe and I now have a belief that we can truly do something significant in the restoration of the two largest flying birds of North America. There are still many hurdles to jump. The trumpeter swan is three times larger than a goose. Similarly, it seems three times as awkward in the air, making flying with it far riskier. From hatching to first flights, swans take over twice as long to begin flying, and winter snows are almost on the ground before they are ready to attempt migration. With swans and cranes there is also still a great deal of politics to consider, for some believe the Trumpeter never existed east of the Mississippi. Whooping cranes imprint in quite a different way than geese; they also fly cross-country with a different technique. Imprinting them on a human can be dangerous, for some experts believe that imprinted birds, when they reach adulthood, consider humans another crane invading their territory and will attack. There are many problems yet unseen and our journey is only beginning, but someday, echoing my first flight, we could be caught up in an equally wondrous multitude of whooping cranes or trumpeter swans!